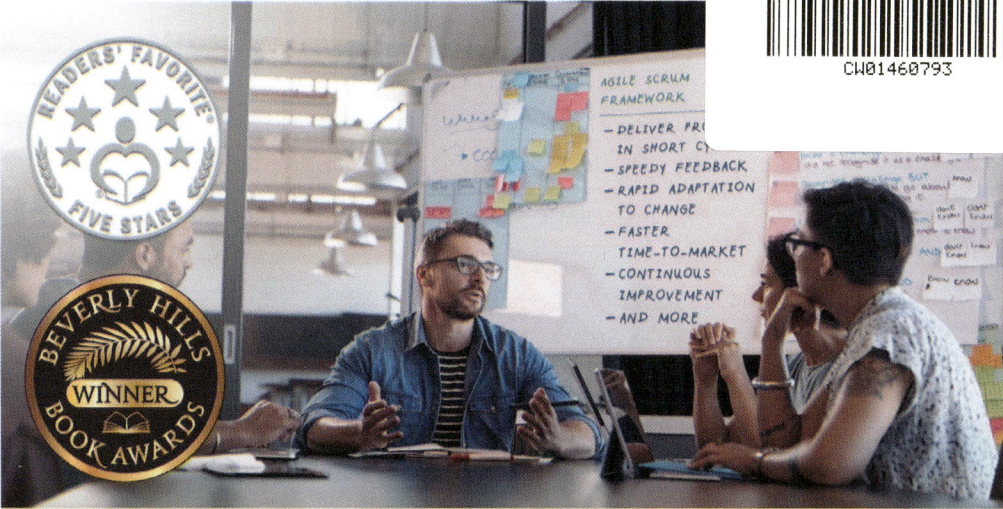

AGILE SCRUM

YOUR QUICK START GUIDE WITH STEP-BY-STEP INSTRUCTIONS

**Deliver Products in Short Cycles
with Rapid Adaptation to Change,
Fast Time-to-Market,
and Continuous Improvement —
Which Supports Innovation
and Drives Competitive Advantage**

SCOTT M. GRAFFIUS

(This page is intentionally blank.)

Praise for
Agile Scrum: Your Quick Start Guide with Step-by-Step Instructions

"A superbly written and presented guide to team-based project management that is applicable across a broad range of businesses from consumer products to high-tech."

— IndieBRAG

"*Agile Scrum: Your Quick Start Guide with Step-by-Step Instructions* is an all-inclusive instruction guide that is impressively 'user-friendly' in tone, content, clarity, organization, and presentation."

— Midwest Book Review

"A-type personalities (those inclined to avoid instruction manuals) and non-readers will rejoice upon discovering this guide which makes getting started with Agile Scrum a breeze."

— Literary Classics Book Reviews

"*Agile Scrum: Your Quick Start Guide with Step-by-Step Instructions* is a must-have for a project manager wanting to introduce Scrum to the organization."

— PM World Journal

"Recommended."

— The US Review of Books

"A clear and authoritative roadmap for successful implementation, *Agile Scrum: Your Quick Start Guide with Step-by-Step Instructions* is strongly recommended."

— BookViral

"★★★★★"

— Readers' Favorite

"★★★★★"

— Amazon.com Hall of Fame and Top 100 Reviewer

More praise for
Agile Scrum: Your Quick Start Guide with Step-by-Step Instructions

Honors include 15 first place awards from national and international competitions.

1st Place Winner
Business-General
5th Annual Beverly Hills
Int'l Book Awards

1st Place Winner
Technology
5th Annual Beverly Hills
Int'l Book Awards

1st Place Winner
Business
2016 London
Book Festival

1st Place Winner
Business
Fall 2016 Pinnacle Book
Achievement Awards

1st Place Winner
Informational
2017 Feathered Quill
Book Awards

1st Place Winner
Technology
2016 New Apple
Book Awards

1st Place Winner
Technology
2017 Independent
Press Award

1st Place Winner
Technology
11th Annual National
Indie Excellence Awards

1st Place Winner
Business
2017 Pacific Rim
Book Festival

1st Place Winner
Green/Conscious
Business
2017 Bookvana Awards

1st Place Winner
Technology
2017 Book Excellence
Awards

1st Place Winner
Business Reference
14th Annual
Best Book Awards

1st Place Winner
Technology
2017 New York City
Big Book Awards

1st Place Winner
Science & Technology
2017 Royal Dragonfly
Book Awards

1st Place Winner
Workplace
2017 Human Relations
Indie Book Awards

Additional reviews and awards are listed at AgileScrumGuide.com.

Agile Scrum: Your Quick Start Guide with Step-by-Step Instructions

Scott M. Graffius

"The progressive development of man is vitally dependent on invention."
— Nikola Tesla
Inventor, electrical engineer, mechanical engineer, physicist, and futurist

---

Book: Graffius, S. M. (2016). *Agile Scrum: Your Quick Start Guide with Step-by-Step Instructions*. North Charleston, SC: CreateSpace.
ISBN-10: 1533370249 | ISBN-13: 978-1533370242
Publisher: CreateSpace, 4900 LaCross Road, North Charleston, SC 29406

eBook: Graffius, S. M. (2016): *Agile Scrum: Your Quick Start Guide with Step-by-Step Instructions*. Seattle, WA: Amazon Digital Services.
ASIN: B01FZOJIIY
Publisher: Amazon Digital Services, 410 Terry Avenue North, Seattle, WA 98109

Version 16040503.18031103

Acknowledgments

Organizations need to realize their strategic objectives. I founded and am CEO of Exceptional PPM and PMO Solutions™, a consultancy which helps our clients achieve their business needs through world-class project leadership. I had the privilege to work with a client in a division of a global entertainment business on their successful journey for improved responsiveness to changing business needs, faster delivery speed, higher satisfaction, and continuous improvement—which made them even more competitive and fueled their growth. That fantastic agile transformation experience was the inspiration for this book. I thank that client.

I am thankful for all of the disruptors, innovators, and visionaries who contributed to the colorful heritage of agile and Scrum. In particular, I thank Hirotaka Takeuchi and Ikujiro Nonaka. Scrum was modeled after their groundbreaking paper, "The New New Product Development Game," published in the *Harvard Business Review* in 1986. This book was informed by their article and 115 additional sources—listed in the bibliography—along with my first-hand experience launching Scrum in organizations.

I owe a tremendous debt to the technical editors: Chris Hare and Colin Giffen. Each offered insights that greatly improved *Agile Scrum: Your Quick Start Guide with Step-by-Step Instructions*. I thank them for helping to make this book more clear, consistent, and valuable.

(This page is intentionally blank.)

About the Author

Scott M. Graffius is a technology leader, project management expert, consultant, international speaker, and author. He is an authority on transforming organizations' ideas and strategic objectives into reality through world-class project, program, portfolio, and PMO management inclusive of agile, traditional, and hybrid frameworks.

Scott is the Founder and CEO of Exceptional PPM and PMO Solutions™, an Inc. Verified Business, which helps organizations achieve their strategic objectives and boost their business results through exceptional project management. The consultancy leverages its expertise, industry standards, and best practices to deliver significant results. Examples include strategically aligned initiatives, faster time-to-market, improved on-budget delivery, higher customer and stakeholder satisfaction, and greater competitive advantage.

He is a former vice president of a provider of diverse consumer products and services over the Internet including social networking and internet access. Before that, Scott worked in organizations with businesses ranging from advanced technology products and services to business services, retail, e-commerce, manufacturing, and entertainment. He has experience with consumer, business, reseller, government, and international customer markets, as well as international experience spanning 20 countries.

Scott holds a bachelor's degree in Psychology with a focus in Human Factors. His certifications include Certified Scrum Professional® (CSP®), Certified ScrumMaster® (CSM®), Certified Scrum Product Owner® (CSPO®), Project Management Professional® (PMP®), IT Service Management Foundation (ITIL®), and Lean Six Sigma Green Belt (LSSGB). He is a member of the Scrum Alliance®, Project Management Institute® (PMI®), Institute of Electrical and Electronics Engineers® (IEEE®), IEEE® Computer Society, IEEE® Consumer Electronics Society, IEEE® Broadcast Technology Society, IEEE® Internet of Things Community, and the IEEE® Consultants Network.

Scott resides in Los Angeles, California.

✉ Scott@ScottGraffius.com

🐦 https://twitter.com/ScottGraffius

www http://ScottGraffius.com

🔊 http://ScottGraffius.com/blog

in https://www.linkedin.com/in/ScottGraffius

BONUS Page 137 has information on how you can provide feedback, access online bonus content, and more

(This page is intentionally blank.)

Table of Contents

Introduction

Welcome

Welcome to *Agile Scrum: Your Quick Start Guide with Step-by-Step Instructions,* your playbook for successfully managing projects and meeting business needs at astounding speed.

There are a variety of frameworks supporting the development of products and services, and most methodologies fall into one of two broad categories: traditional or agile. Traditional practices engage sequential development, while agile involves iterative and incremental deliverables. Organizations are increasingly embracing agile to best meet their business needs and effectively manage projects.

With clear and easy to follow step-by-step instructions, this guide helps you:

- Implement and use the most popular agile framework—Scrum
- Deliver products in short cycles with rapid adaptation to change, fast time-to-market, and continuous improvement
- Support innovation and drive competitive advantage

Audience

This guide is for those interested or involved in innovation, project management, product development, software development or technology management.

- It's for those who have not yet used Scrum
- It's also for people already using Scrum, in roles such as Product Owners, Scrum Masters, Development Team members (business analysts, solution and system architects, designers, developers, testers, etc.), customers, end users, agile coaches, executives, managers, and other stakeholders

For those already using Scrum, this book can serve as a reference on practices for consideration and potential adaptation.

Sections in this Book

Here's how this guide is organized:

- A quick introduction
- The main section provides the how-to information with answers to who, what, when, where, how, and how much
- Additional content presents further details and answers why the steps in the main section are important; it includes background information, a glossary, examples, illustrations, tips, a project assessment tool, and more

Supplemental Information

Supplemental information indicators let you know that there's additional related information in the appendix of this book. The tags (such as "Agile Scrum Overview") tell you the sections of the appendix where the information is located. You can skip them if you'd like, but the extra content is informative, concise (usually about a page per topic) and there for your reference whenever you need it.

If you are interested in background information on Agile Scrum, you'll find it—and more—in the appendix.

> There's supplemental information in the following sections of the appendix:
>
> Agile Scrum Overview   Agile Manifesto   Agile Principles
>
> Inspect and Adapt

If you would like to see how Scrum differs from traditional development and delivery methods, there are side-by-side comparisons in the appendix.

> There's supplemental information in the following sections of the appendix:
>
> Communications Management Comparison   Cost Management Comparison
>
> Human Resources/Team Management Comparison
>
> Integration Management Comparison   Quality Management Comparison
>
> Risk Management Comparison   Scope Management Comparison
>
> Time Management Comparison

That's it for the brief introduction. Let's move on to the instructions.

How to Easily Employ Agile Scrum (Step-by-Step Instructions)

Overview

This guide provides you with instructions for the successful implementation and use of Agile Scrum, and the image below presents a high-level view of what's involved.

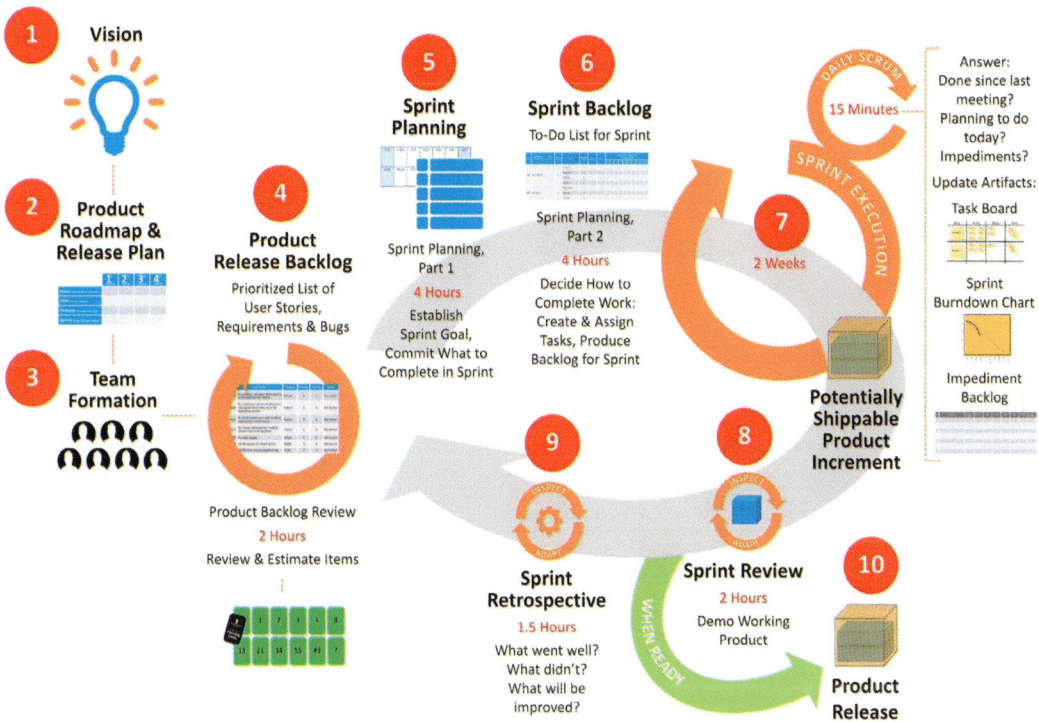

Agile Scrum Cycle

Time-boxes Shown are Based on the Most Common Sprint Duration of Two Weeks

© Scott M. Graffius

The Agile Scrum framework has been sliced into ten sections to make the information easier to read, understand and use. The instructions take you through the framework one section at a time.

This guide incorporates the values and principles of agile along with information on typical successful Scrum implementations and best practices. For example, velocity—the most popular metric—is not used by some people. However, this guide covers velocity and other leading practices.

Agile Scrum is not an all-or-nothing proposition, and every aspect does not need to be followed to the letter. Just as a value of agile is "individuals and interactions over processes and tools" and a principle is "inspect and adapt," whatever approach you decide to use should be adjusted as appropriate to your unique circumstances.

There's supplemental information in the following sections of the appendix:

Agile Manifesto   Agile Principles   Inspect and Adapt

When to Use Agile Scrum

Up next is **Section 1: Vision**.

Section 1: Vision

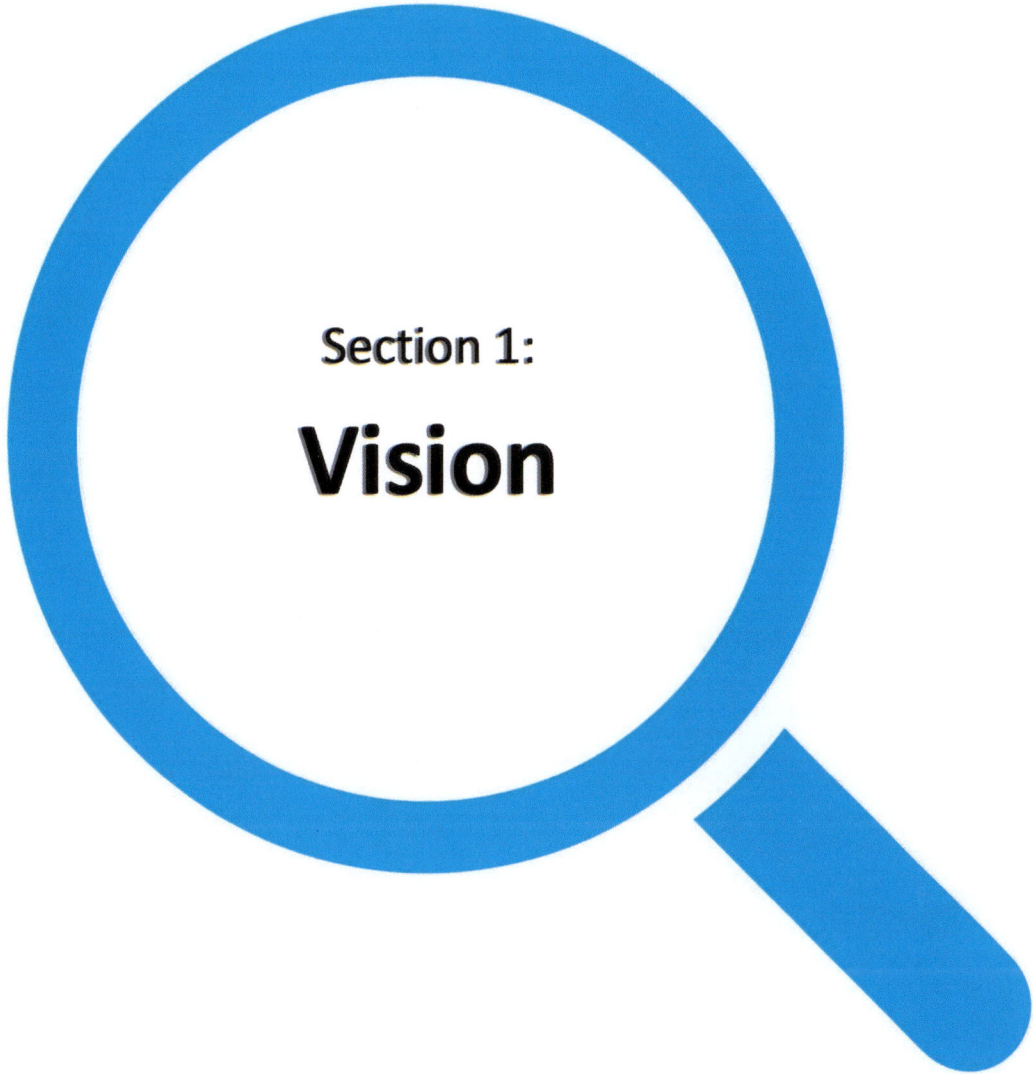

Section 1:

Vision

Agile Scrum Cycle

Time-boxes Shown are Based on the Most Common Sprint Duration of Two Weeks
© Scott M. Graffius

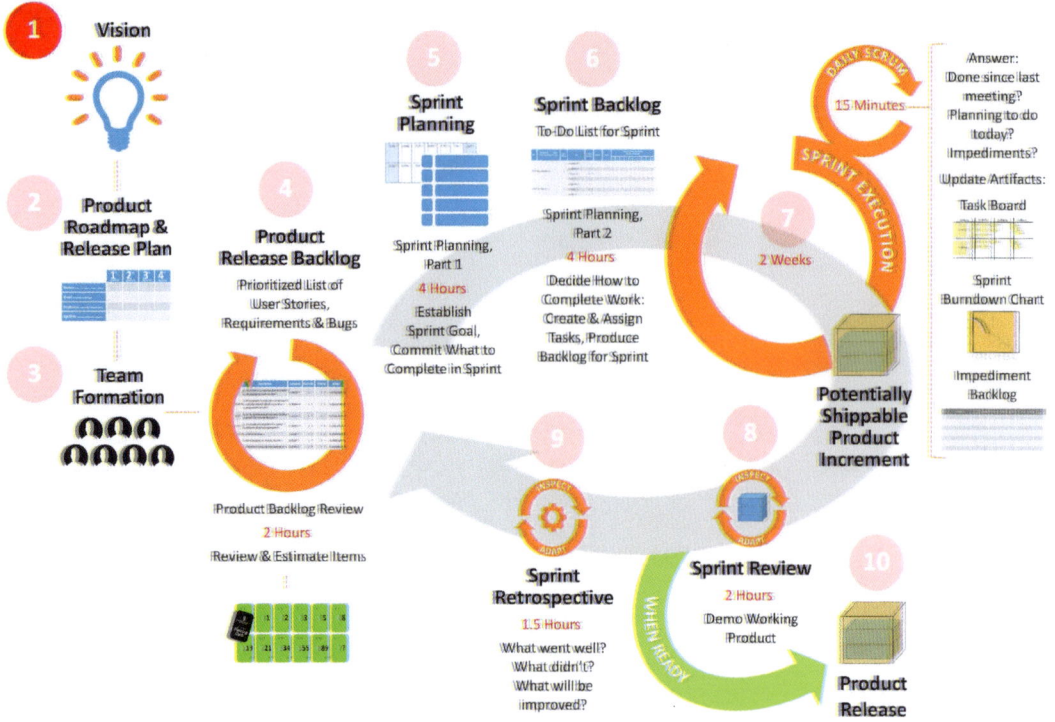

1 Vision

2 Product Roadmap & Release Plan

3 Team Formation

4 Product Release Backlog
Prioritized List of User Stories, Requirements & Bugs

Product Backlog Review
2 Hours
Review & Estimate Items

5 Sprint Planning

Sprint Planning, Part 1
4 Hours
Establish Sprint Goal, Commit What to Complete in Sprint

6 Sprint Backlog
To-Do List for Sprint

Sprint Planning, Part 2
4 Hours
Decide How to Complete Work: Create & Assign Tasks, Produce Backlog for Sprint

DAILY SCRUM
15 Minutes

SPRINT EXECUTION
7 2 Weeks

Answer:
Done since last meeting?
Planning to do today?
Impediments?

Update Artifacts:
Task Board

Sprint Burndown Chart

Impediment Backlog

Potentially Shippable Product Increment

8

9 Sprint Retrospective
1.5 Hours
What went well?
What didn't?
What will be improved?

WHEN READY

8 Sprint Review
2 Hours
Demo Working Product

10 Product Release

Purpose

The product vision is the overall objective. It's created by the Product Owner, and it describes in about two sentences the product's purpose, who it is for, how it will create value, and the benefits. The product vision is sometimes called the product goal.

This topic—**Section 1: Vision**—is labeled with **1** in the illustration above.

There's supplemental information in the following section of the appendix:

Product Owner

Participants

- Product Owner

- Stakeholders (for feedback)

- Scrum Master (as applicable, for feedback)

- Development Team (as applicable, for feedback)

> There's supplemental information in the following sections of the appendix:
>
> | Product Owner | Stakeholders | Scrum Master | Development Team |

Frequency or sequence

- Typically, once per year

- It may be done more often such as semiannually or quarterly

Time-box (not-to-exceed duration)

There's no time-box for this activity

Steps

1. The Product Owner answers the following questions:

   - How will the product benefit the company?

   - Who will be the customer or end user?

   - How will the product benefit the customer or end user?

   - How is this product similar to existing products?

   - How is this product different than existing products?

   - What is the timeframe for the product?

   - What is the budget to create and implement the product?

2. The Product Owner drafts the vision statement

- A template by Geoffrey Moore may be helpful. Here's an overview:

    For: <customer or end user>

    Who: <needs>

    The: <product name>

    Is a: <product category>

    That: <product benefit, or reason to buy or use it>

    Unlike: <competitors or alternatives>

    Our product: <differentiation or value proposition>

- Example:

    "For small size companies, who need a simple business intelligence tool, Widgets is a cloud-based offering that provides a process for analyzing data and presenting actionable information to help executives make well-informed business decisions. Unlike similar services, our product works offline"

3. The Product Owner sends a draft of the vision statement with a request for feedback to:

    - Stakeholders

    - Scrum Master (if already assigned/involved)

    - Development Team (if already assigned/involved)

4. The Product Owner receives feedback

5. The Product Owner revises and finalizes the vision statement

6. The Product Owner circulates the final vision statement to those involved in step 3

7. The Product Owner posts the final vision statement at a prominent location where the Scrum team and stakeholders can see it

Up next is **Section 2: Product Roadmap and Release Plan**.

Section 2:

Product Roadmap and Release Plan

Agile Scrum Cycle

Time-boxes Shown are Based on the Most Common Sprint Duration of Two Weeks
© Scott M. Graffius

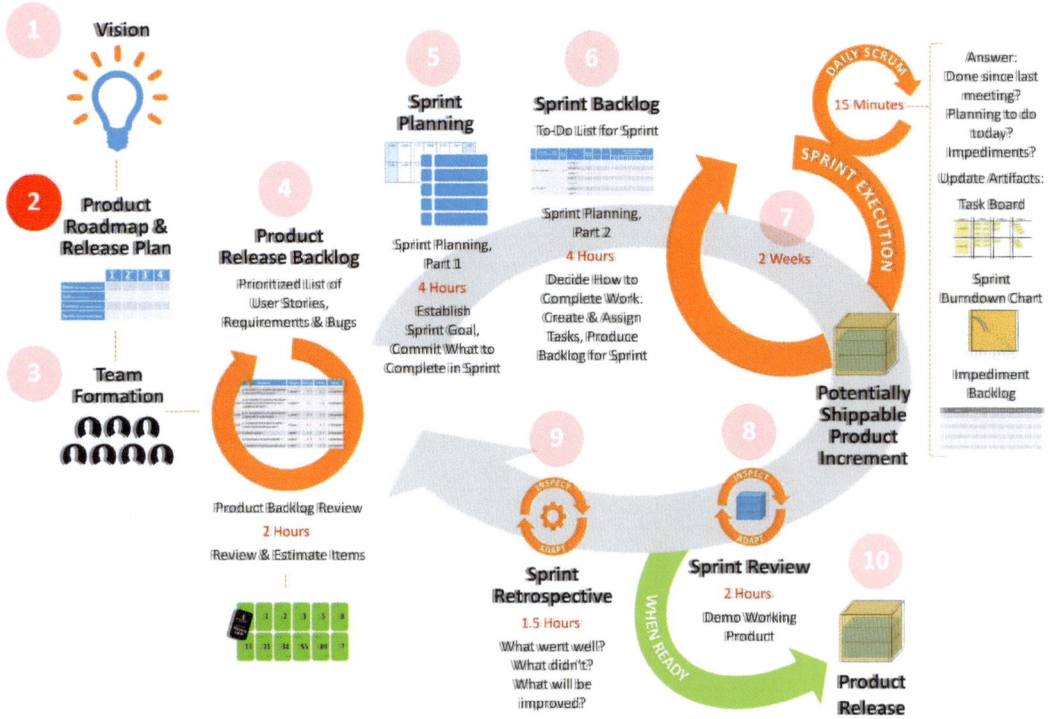

1 Vision

2 Product Roadmap & Release Plan

3 Team Formation

4 Product Release Backlog
Prioritized List of User Stories, Requirements & Bugs

Product Backlog Review
2 Hours
Review & Estimate Items

5 Sprint Planning
Sprint Planning, Part 1
4 Hours
Establish Sprint Goal, Commit What to Complete in Sprint

6 Sprint Backlog
To-Do List for Sprint
Sprint Planning, Part 2
4 Hours
Decide How to Complete Work: Create & Assign Tasks, Produce Backlog for Sprint

DAILY SCRUM
15 Minutes

SPRINT EXECUTION

7 2 Weeks

Answer: Done since last meeting? Planning to do today? Impediments?
Update Artifacts:
Task Board
Sprint Burndown Chart
Impediment Backlog

8 Potentially Shippable Product Increment
INSPECT ADAPT

9 Sprint Retrospective
1.5 Hours
What went well? What didn't? What will be improved?
INSPECT ADAPT

Sprint Review
2 Hours
Demo Working Product

WHEN READY

10 Product Release

Purpose

The combination product roadmap and release plan is a high-level description of how a product will be built and deployed over time.

This topic—**Section 2: Product Roadmap and Release Plan**—is tagged with **2** in the illustration above.

Participants

- Product Owner

- Stakeholders (for feedback)

- Scrum Master (as applicable, for feedback)

- Development Team (as applicable, for feedback)

There's supplemental information in the following sections of the appendix:

| Product Owner | | Scrum Master | | Development Team |

| Stakeholders |

Frequency or sequence

- Typically, twice per year

- It may be done more often such as quarterly

Time-box (not-to-exceed duration)

There's no time-box for this activity

Prerequisites or inputs

Completion of **Section 1: Vision**

Steps

1. The Product Owner drafts the product roadmap and release plan as an extension and elaboration of the product vision

   - Product roadmap and release plans come in different forms, but they typically involve a mini-plan for each quarter with the following information:

     o Name

       ▪ Title for the product or major release

       ▪ Examples:

         – Dance Game 1

           o In addition to or instead of a high-level name (Dance Game 1 in this case), a code name or version may be used

             ▪ Code name: Electric Puma

             ▪ Version 1.0

  - Diagramming Tool 2016

    o In addition to or instead of a high-level name (Diagramming Tool 2016 in this case), a code name or version may be used

      ▪ Code name: Transformer

      ▪ Version 16.31

o Goal

  ▪ The reason for creating it

  ▪ Examples:

    - Dance Game 1

      o Product introduction

      o Customer acquisition

    - Diagramming Tool 2016

      o Customer acquisition

      o Customer retention

o Features

  ▪ A brief, high-level list of features

  ▪ Examples:

    - Dance Game 1

      o Single and multi-player game

      o Featuring 21 songs

      o 650 dance moves with 80 dance routines

      o Three modes: perform, workout, and challenge

    - Diagramming Tool 2016

- o Add 100 new starter diagrams

- o Contextual tips and tricks throughout

- o Support simultaneous co-authoring

- o Offer annual subscription option

- o Sprints

  - A forecast of the number of sprints

    - Information on sprints may be left blank initially

    - It can be updated after sprints have been completed

      - o Here's how to calculate it: Divide the total story points for the features in the release by the team's average velocity per sprint

      - o Story points are covered in **Section 4: Product Release Backlog**, and velocity is covered in **Section 7: Sprint Execution** and additional parts of this guide

- A sample template for the product roadmap and release plan is shown below

Product Roadmap and Release Plan

				1ST QUARTER	2ND QUARTER	3RD QUARTER	4TH QUARTER
	NAME	Title of Product or Major Release					
	GOAL	Reason for Creating It					
	FEATURES	List of High-Level Features					
	SPRINTS	Forecast of Number of Sprints					

- Additional considerations:

  o Some product roadmaps and release plans include metrics

    ▪ Examples:

      – Number of new customers/subscribers

      – Number of page views

      – Number of downloads

      – Or whatever other measures are applicable

  o Other supplemental information can be incorporated

  o If you're new to Scrum, you may find it best to begin with the use of flip chart paper, a whiteboard or a simple spreadsheet

  o Specialized software tools—such as those listed in **Select Resources**—can always be used later

2. The Product Owner sends a draft of the product roadmap and release plan with a request for feedback to:

   - Stakeholders

   - Scrum Master (if already assigned/involved)

   - Development Team (if already assigned/involved)

3. The Product Owner receives feedback

4. The Product Owner revises the product roadmap and release plan

5. The Product Owner circulates the updated product roadmap and release plan to those involved in step 2 above

6. The Product Owner posts the product roadmap and release plan at a prominent location where the Scrum team and stakeholders can see it

The preceding instructions focused on the initial creation of a product roadmap and release plan. It's a dynamic document; and whenever an update is needed, the Product Owner edits it and then repeats steps 2-6.

Up next is **Section 3: Team Formation**.

Section 3: Team Formation

Section 3:

Team Formation

Agile Scrum Cycle

Time-boxes Shown are Based on the Most Common Sprint Duration of Two Weeks
© Scott M. Graffius

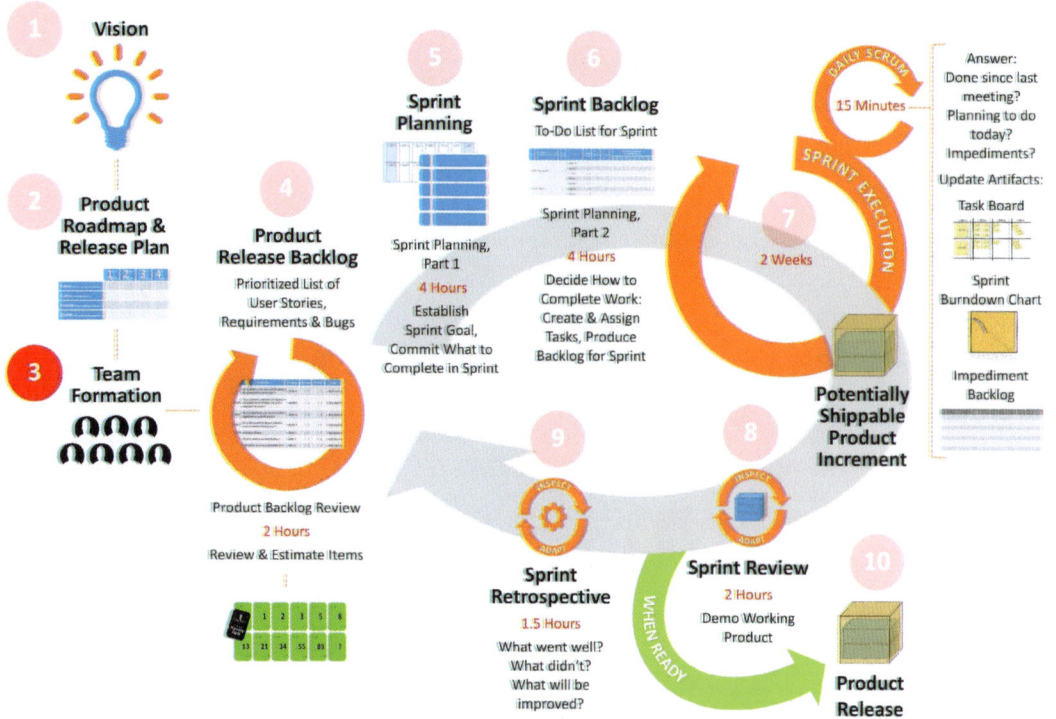

1 Vision

2 Product Roadmap & Release Plan

3 Team Formation

4 Product Release Backlog
Prioritized List of User Stories, Requirements & Bugs

Product Backlog Review
2 Hours
Review & Estimate Items

5 Sprint Planning
Sprint Planning, Part 1
4 Hours
Establish Sprint Goal, Commit What to Complete in Sprint

6 Sprint Backlog
To-Do List for Sprint
Sprint Planning, Part 2
4 Hours
Decide How to Complete Work: Create & Assign Tasks, Produce Backlog for Sprint

7 2 Weeks

DAILY SCRUM
15 Minutes
Answer: Done since last meeting? Planning to do today? Impediments?
Update Artifacts: Task Board, Sprint Burndown Chart, Impediment Backlog

SPRINT EXECUTION

Potentially Shippable Product Increment

8 Sprint Review
2 Hours
Demo Working Product

9 Sprint Retrospective
1.5 Hours
What went well? What didn't? What will be improved?

WHEN READY

10 Product Release

Purpose

During team formation, the full Scrum team (Product Owner, Scrum Master, and Development Team) joins the project if they're not already engaged in it.

This topic—**Section 3: Team Formation**—is labeled with **3** in the illustration above.

Participants

- Product Owner

- Scrum Master

- Development Team

- (Another individual such as a member of management may be involved—this is explained later in this section)

There's supplemental information in the following sections of the appendix:

Product Owner   Scrum Master   Development Team

Frequency or sequence

Team formation is visited as needed before work starts in the next phase, **Section 4: Product Release Backlog**

Time-box (not-to-exceed duration)

There's no time-box for this activity

Prerequisites or inputs

- Typically, completion of **Section 1: Vision**

- Generally, completion of **Section 2: Product Roadmap and Release Plan**

Steps

1. If the full Scrum team is already assigned, you can advance to the next phase (**Section 4: Product Release Backlog**). If the full Scrum team is not yet assigned, here are some guidelines:

   - The full Scrum team totals between five and eleven (formerly between seven and eleven) people, comprised of:

     o One Product Owner

     o One Scrum Master

     o Between three and nine (formerly between five and nine) Development Team members—business analysts, coders, testers, etc.

   - Favorable characteristics for the team include:

     o Self-organizing

     o Cross-functional

- o   Collaborative

- o   Competent

- o   Dedicated—100% assigned to the project

- o   Experienced

- o   Motivated

- •   Additional factors to be considered when assembling the team include:

  - o   Expertise

  - o   Planned absences and other commitments

2.   The appropriate person assembles the team

- •   The responsible role varies by organization—the person might be the:

  - o   Product Owner

  - o   Scrum Master

  - o   A member of management

  - o   Or another person

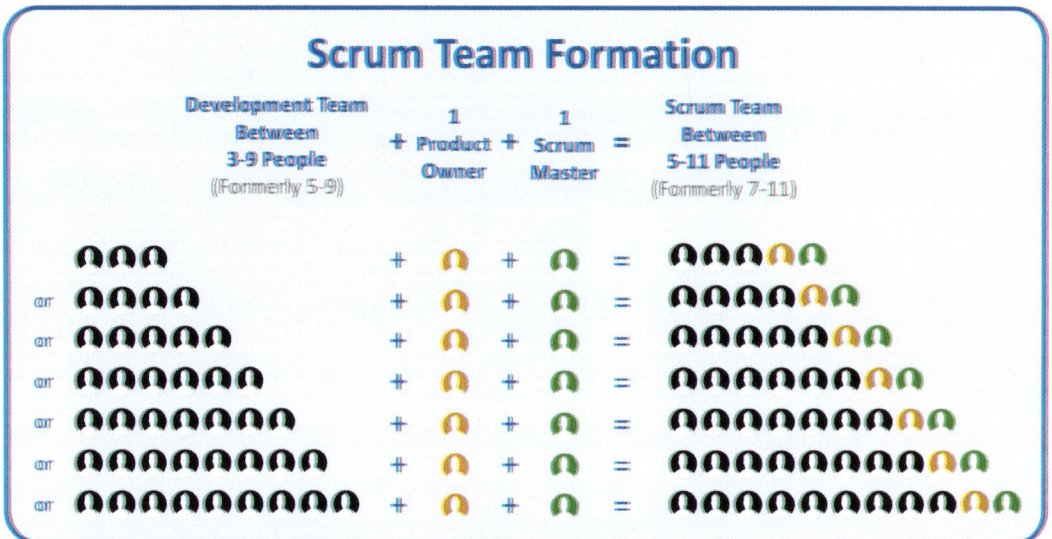

Scrum Team Formation

Development Team Between 3-9 People ((Formerly 5-9))		1 Product Owner		1 Scrum Master		Scrum Team Between 5-11 People ((Formerly 7-11))

Up next is **Section 4: Product Release Backlog.**

Section 4: Product Release Backlog

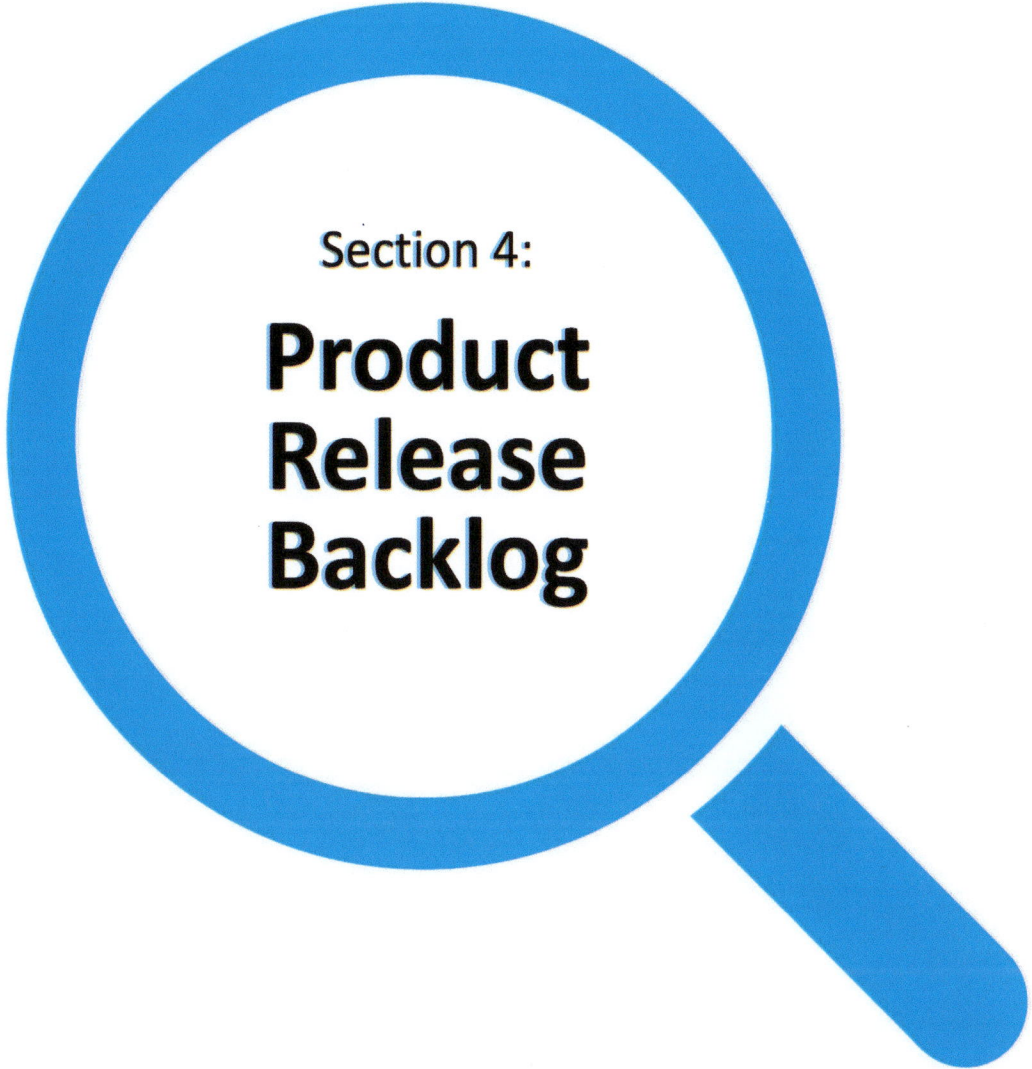

Section 4:

Product Release Backlog

Agile Scrum Cycle

Time-boxes Shown are Based on the Most Common Sprint Duration of Two Weeks
© Scott M. Graffius

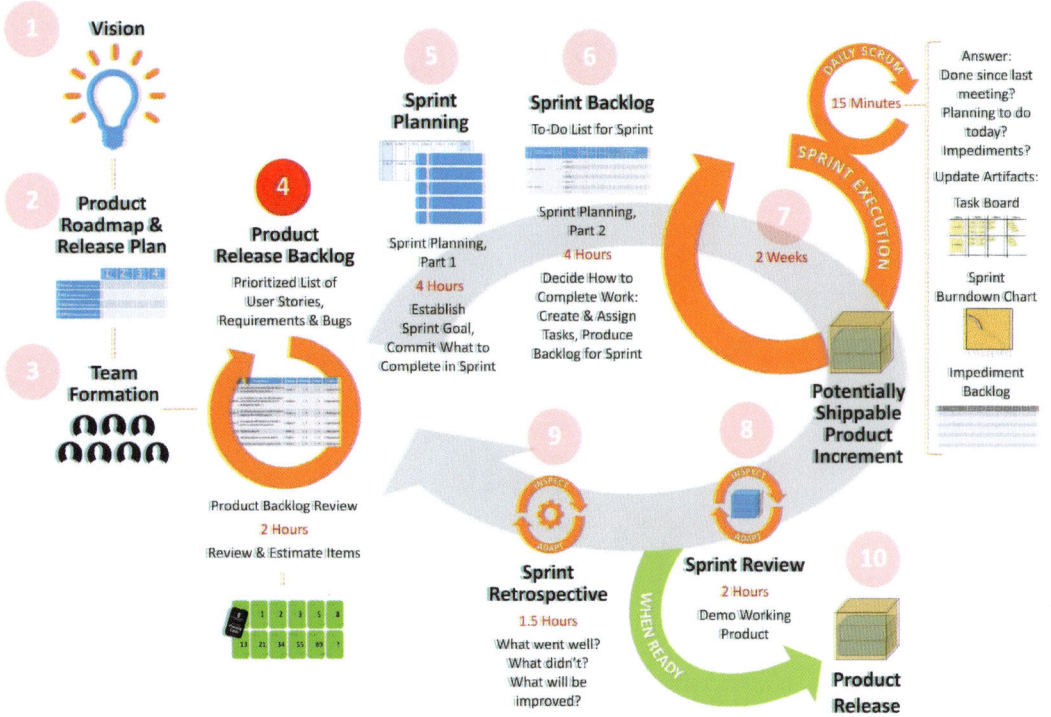

1 Vision

2 Product Roadmap & Release Plan

3 Team Formation

4 Product Release Backlog
Prioritized List of User Stories, Requirements & Bugs

Product Backlog Review
2 Hours
Review & Estimate Items

5 Sprint Planning

Sprint Planning, Part 1
4 Hours
Establish Sprint Goal, Commit What to Complete in Sprint

6 Sprint Backlog
To-Do List for Sprint

Sprint Planning, Part 2
4 Hours
Decide How to Complete Work: Create & Assign Tasks, Produce Backlog for Sprint

DAILY SCRUM
15 Minutes

SPRINT EXECUTION

7 2 Weeks

Answer:
Done since last meeting?
Planning to do today?
Impediments?
Update Artifacts:
Task Board
Sprint Burndown Chart
Impediment Backlog

Potentially Shippable Product Increment

9 Sprint Retrospective
1.5 Hours
What went well?
What didn't?
What will be improved?

8 Sprint Review
2 Hours
Demo Working Product

WHEN READY

10 Product Release

This topic—**Section 4: Product Release Backlog**—is tagged with **4** in the illustration above.

Quick note

To make the Product Release Backlog material easier to follow, there are two subsections:

- **4A: Product Release Backlog with User Stories and Prioritized Entries**

- **4B: Estimation via Story Points**

4A: Product Release Backlog with User Stories and Prioritized Entries

Purpose

The product release backlog is established. It's a list of features, requirements, and bugs— all prioritized by value. It's also known as the product backlog and the release backlog.

Participants

Product Owner

> There's supplemental information in the following section of the appendix:
>
> Product Owner

Frequency or sequence

Variable frequency/ongoing

Time-box (not-to-exceed duration)

There's no time-box for this activity

Prerequisites or inputs

- Completion of **Section 1: Vision**

- Completion of **Section 2: Product Roadmap and Release Plan**

Steps

1. The Product Owner creates the product release backlog:

   - Product release backlogs take on different forms, ranging from information hand-drawn on flip chart paper or whiteboards to the use of software tools

     o If you are new to Scrum, you may find it best to begin with the use of flip chart paper or a whiteboard

     o Specialized software tools—such as those listed in **Select Resources**—can always be used later

   - Each entry in the product release backlog is typically comprised of (at least) the following:

- o   ID number

- o   Description including acceptance criteria

- o   Category

- o   Estimate in story points (or an alternative method)

- o   Priority

- o   Status

- A product release backlog sample template is shown below

Product Release Backlog

ID #	Description	Category	Story Points	Priority	Status

2. The Product Owner creates a user story description for each item and adds it to the product release backlog (note: some organizations do not write user stories for bugs)

- The user story description represents a portion of business value that a team can deliver in an iteration

- The focus is on the value that a user gains from the system

- User stories are written in plain language

- A common format of a user story is:

   - o   As a <role>, I want <goal> so that I can <reason>

   - o   Example: "As a customer, I want shopping cart functionality so that I can buy items online"

- A well-written user story (sometimes called a product backlog item) follows the INVEST model developed by Bill Wake:

  - INVEST stands for:

    - Independent: Stories are self-contained, free of dependencies

    - Negotiable: Stories capture the essence and leave some room for discussion

    - Valuable: Stories and their follow-on work products deliver value

    - Estimable: Stories can be estimated

    - Small: Stories tend to be small in size, such as amounting to a few days of work (note: some organizations limit the size of work to what can be done in no more than one workday)

    - Testable: Stories must provide information sufficient to make them testable

There's supplemental information in the following section of the appendix:

User Stories—Techniques for Gathering Requirements

3. The Product Owner creates acceptance criteria for each item and adds it to the product release backlog (as part of the description, or in an additional column)

- Guidelines and examples:

  - Acceptance criteria are conditions that a product must satisfy to be accepted by a user, customer or another stakeholder

  - They're requirements that must be met

  - They add certainty to what the team is building

  - Acceptance criteria must be expressed clearly in plain language, and they must be testable

  - They may cover functional, non-functional or performance criteria

    - Functional criteria

- Specific user tasks, functions or business processes

- Example: "A user can access logs"

- Non-functional criteria

  - Specific non-functional conditions the implementation must meet

  - Example: "Print buttons comply with the button design"

  - To support quality standards, the following non-functional criteria should be factored:

    - Documented

    - Extensible

    - Maintainable

    - Modular

    - Portable

    - Reliable

    - Resilient

    - Reusable

    - Robust

    - Scalable

    - Secure

    - Sufficient

    - Testable

- Performance criteria

  - Specific performance critical to the acceptance of a user story, if applicable

    - Example: "No longer than 1.00 second for results returned to user"

4. The Product Owner notes the category for the item

   - Common categories are:

     o Feature

     o Requirement

     o Bug

5. The Product Owner enters the estimates in story points (or an alternative method) later

   - It's covered in **4B: Estimation via Story Points**

6. The Product Owner prioritizes each item and adds that information to the product release backlog

   - "MoSCoW" is a popular acronym that represents a method of ranking stories

     o With "MoSCoW," each item is sorted into one of four categories:

        ▪ "Must have", "Should have", "Could have", and "Won't have"

     o Backlog items rated as "Must have" are the most important

        ▪ Identify the "Must have" items and further prioritize them with "1" for the top one, "2" for the next one, and so on

        ▪ After addressing the "Must have" items, proceed through the backlog denoting remaining items as either:

           – "Should have," "could have," or "won't have"

              ▪ "Won't have" items won't be done—you may find that half or more of the items fall into this category

   - Another method involves factoring business value and risk

     o Example:

        ▪ Each item in the product release backlog would be rated as high or low in two dimensions—business value and risk

        ▪ It is suggested that high business value, high risk items are worked on first

— The earlier this work is done, the sooner the team will move to mitigate the issues and unknowns, leading to a higher quality product

— If there's a failure, it will occur early and relatively inexpensively

▪ An ordering of priorities is illustrated below

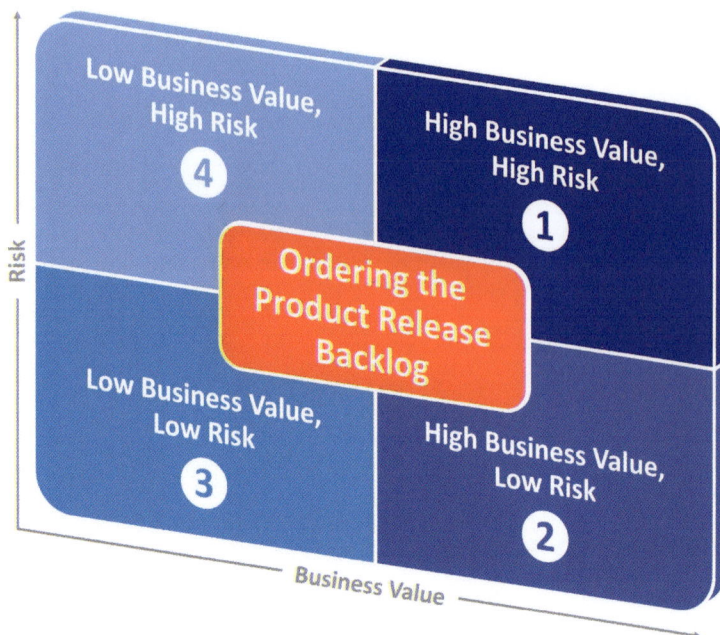

• Other prioritization methods may be used

7. The Product Owner indicates the status of the item

• "Not Committed to a Sprint" or equivalent is the initial status, and the item would be updated to "Committed to Sprint" or similar when it's selected to be done

The preceding instructions focused on the initial creation of a product release backlog. It's a dynamic document; and whenever an update is needed, the Product Owner edits it.

> There's supplemental information in the following section of the appendix:
>
> Technical Debt

Up next is **4B: Estimation via Story Points**.

4B: Estimation via Story Points

Purpose

For planning, the Product Owner obtains high-level initial estimates of the complexity of items in the product release backlog. That information helps to inform decisions.

The Development Team collaboratively estimates each item in the product release backlog in story points. Story points are a relative measure of complexity, not units of time. (Estimates of hours for work occurs when team members are about to start the work—that's covered later.)

The story estimate game—it may also be known by other names—is a popular approach to estimation via story points. The following instructions pertain to that method.

> There's supplemental information in the following section of the appendix:
>
> Fibonacci Sequence and Story Points

Participants

- Product Owner

- Development Team

- Scrum Master (facilitator/observer)

> There's supplemental information in the following sections of the appendix:
>
> Product Owner   Scrum Master   Development Team

Frequency or sequence

- Once per sprint

  o Now

- Or possibly two per sprint

  o Now

  o And another during the sprint (this is explained later)

Time-box (not-to-exceed duration)

- One hour for each week of the sprint

    o   It's a common practice to limit each meeting to one hour and have multiple meetings as appropriate

    o   In the case of the most common sprint duration of two weeks, the total meeting time would be two hours

        ▪   In this example, there would be two separate, one-hour meetings

            –   One now

            –   And one later on during the sprint—the session may be referred to as an estimation meeting

Prerequisites or inputs

- Completion of **4A: Product Release Backlog with User Stories and Entries Prioritized**

- Estimation planning cards (shown in step 1, below)

Steps

1. If each member of the Development Team does not already have their own set of estimation planning cards, the Scrum Master provides materials as needed

    - An example of a set of cards is shown below

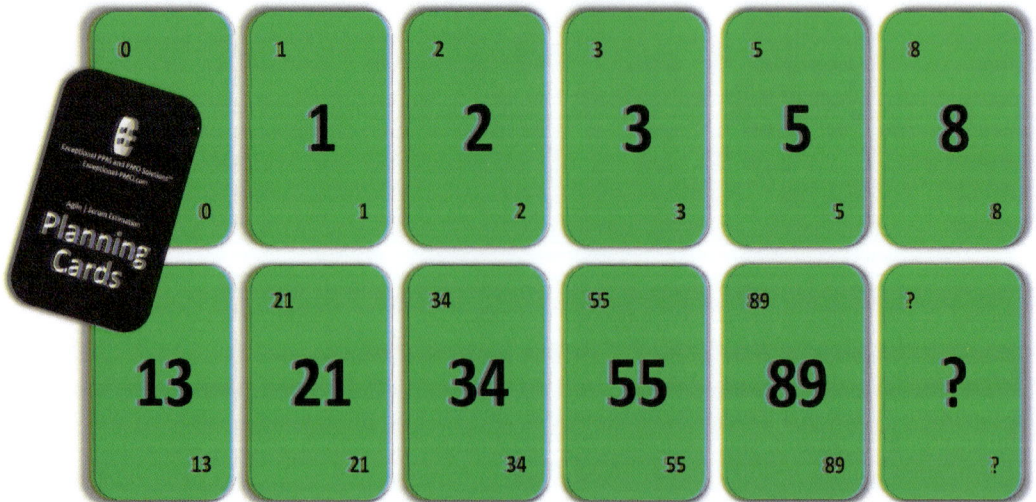

- o   Some organizations use cards with different names or values, but the overall process is usually similar

2.   The Product Owner describes an item—a user story, bug or other requirement—from the product release backlog and mentions its intent and business value

3.   Each member of the Development Team silently picks a card best representing their assessment of the complexity of the work and places the card face-down

- •   Story points are a relative measure of complexity, not units of time

  - o   Estimates for hours of work will be handled later

4.   After all of the Development Team members have made their selections, the cards are turned face-up, and the values are read aloud

5.   If all of the selections have the same value, the Product Owner records it as the estimate, and that completes the exercise for the item; otherwise, proceed to the next step

6.   Team member(s) who gave an outlier value—such as someone who gave a high value and/or someone who gave a low value—explain their reasoning

7.   After a brief discussion, the team may:

- •   Take the most common value (the mode average) from the Development Team as the estimate, or

- •   Play another round for the item (steps 3-7)

8.   Steps 2-7 are repeated until each item in the product release backlog has been estimated

9.   The Product Owner updates the product release backlog with the estimate values

The preceding instructions referenced the use of estimation planning cards based on the Fibonacci Sequence—a popular method for estimation in Scrum—where cards have values ranging from 0 to 89. And there's a "?" card which can be used when a Development Team member is unsure and needs more information in order to provide an estimate.

There's supplemental information in the following section of the appendix:

Fibonacci Sequence and Story Points

Some organizations use a subset of the sequence of estimation planning cards and slice (split) product release backlog items when the estimate is "too large." For example:

- Development Team uses cards with the following values: 0, 1, 2, 3, 5, 8, 13, and 21

- Predetermined that 21 is "too large"

- If the Development Team estimates a product release backlog item at 21, the item is sliced (split) into two or more parts in collaboration with the Product Owner, and the resulting smaller items are estimated by the Development Team

---

More Options for Estimation

Another method involves using t-shirt sizes—such as small, medium, large, and extra-large—as a scale for estimates.

Some teams employ smartphone apps as alternatives to story points or t-shirt sizes on physical cards. Tools are listed in **Select Resources**.

Up next is **Section 5: Sprint Planning Part 1**.

Section 5: Sprint Planning Part 1

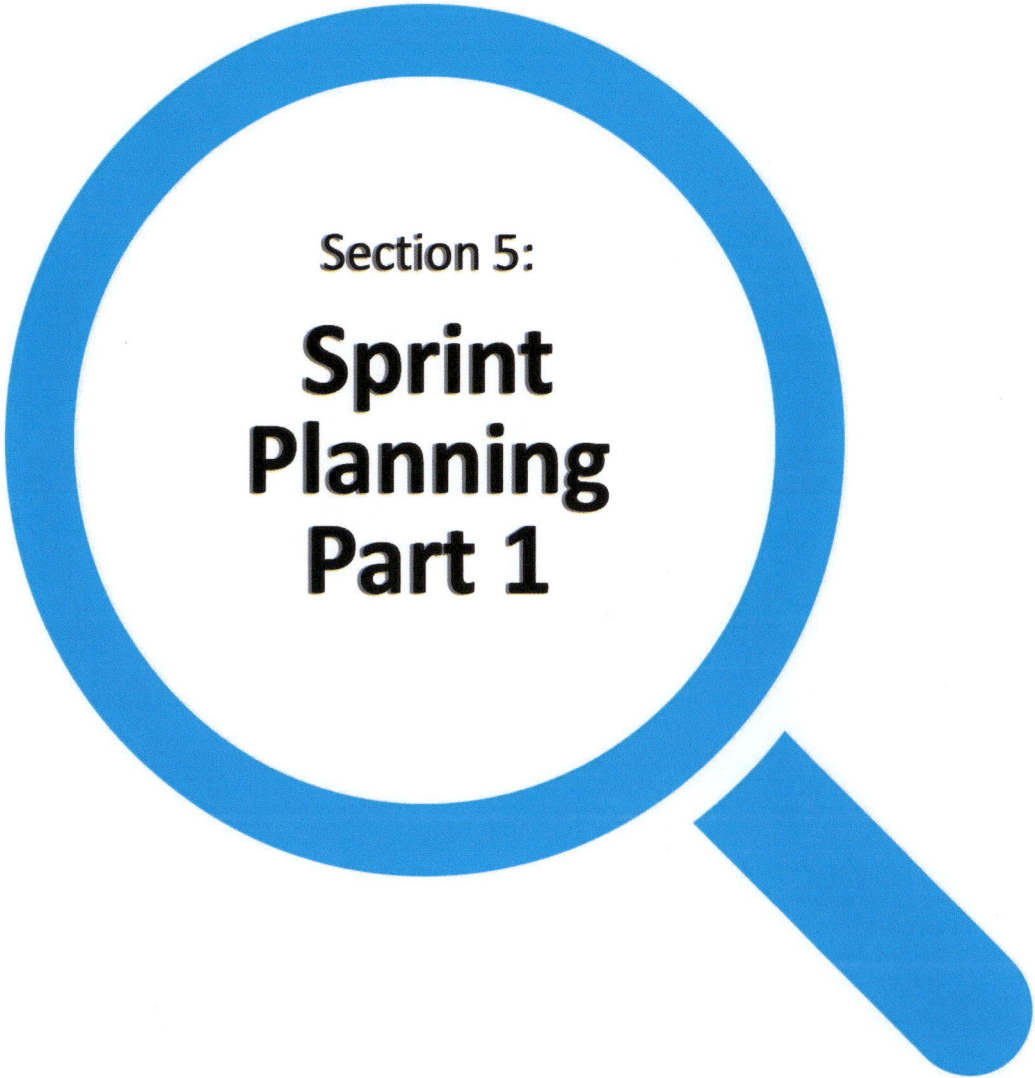

Section 5:

Sprint Planning Part 1

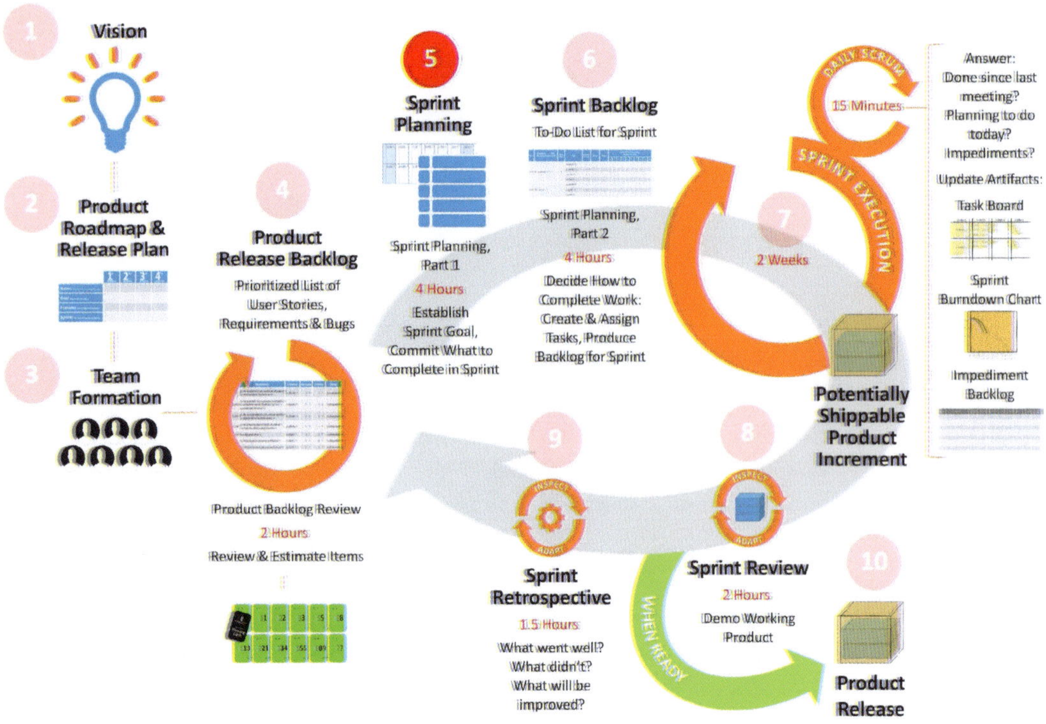

Agile Scrum Cycle

Time-boxes Shown are Based on the Most Common Sprint Duration of Two Weeks
© Scott M. Graffius

1 Vision

2 Product Roadmap & Release Plan

3 Team Formation

4 Product Release Backlog
Prioritized List of User Stories, Requirements & Bugs

Product Backlog Review
2 Hours
Review & Estimate Items

5 Sprint Planning
Sprint Planning, Part 1
4 Hours
Establish Sprint Goal, Commit What to Complete in Sprint

6 Sprint Backlog
To-Do List for Sprint
Sprint Planning, Part 2
4 Hours
Decide How to Complete Work: Create & Assign Tasks, Produce Backlog for Sprint

DAILY SCRUM
15 Minutes

SPRINT EXECUTION

7 2 Weeks

Answer:
Done since last meeting?
Planning to do today?
Impediments?

Update Artifacts:
Task Board

Sprint Burndown Chart

Impediment Backlog

8 Potentially Shippable Product Increment
INSPECT ADAPT

9 Sprint Retrospective
1.5 Hours
What went well?
What didn't?
What will be improved?
INSPECT ADAPT

Sprint Review
2 Hours
Demo Working Product

WHEN READY

10 Product Release

Purpose

During sprint planning, the Development Team pulls items from the top of the product release backlog to create a sprint backlog (part 1—the "what" meeting) and then decides how to accomplish those items during the next sprint (part 2—the "how" meeting). Part 1 is covered in this section. Sprint Planning is referred to as an event (formerly a ceremony).

This topic—**Section 5: Sprint Planning Part 1**—is labeled with **5** in the illustration above.

Participants

- Scrum Master

- Product Owner

- Development Team

There's supplemental information in the following sections of the appendix:

| Product Owner | Scrum Master | Development Team |

Frequency or sequence

Before the start of each sprint

Time-box (not-to-exceed duration)

Time-boxed to two hours for each week of the sprint

Prerequisites or inputs

Completion of **Section 4: Product Release Backlog**

Steps

1. The Scrum Master makes the following information visible during the meeting:

    - Start and end dates for the sprint

    - Results from the last sprint review meeting if the Scrum team has worked together before

    - Results from the last sprint retrospective meeting if the Scrum team has worked together before

2. The Product Owner informs the team about the product vision

    - See **Section 1: Vision**

3. The Product Owner establishes and shares the sprint goal

    - The sprint goal provides a brief description of the objective of the sprint

        o It's usually a single sentence

        o For example:

            ▪ "Implement shopping cart functionality, including add, remove, and update items, in addition to payment processing"

4. The Development Team determines their capacity for the upcoming sprint

- Development Team capacity (in hours) is calculated as the number of Development Team members, multiplied by the number of project productive hours per workday, multiplied by the number of work days in the sprint

  o Project productive hours should exclude time outside the sprint, such as company meetings, trainings, vacation time, etc.

  o As a general guideline, for planning:

    ▪ Subtract 30% as detailed below

      − 5% for company internal meetings and trainings

      − 5% for daily Scrums and follow-ups

      − 10% for sprint planning

      − 3% for sprint demo preparations

      − 7% buffer for unplanned absences such as sick time

    ▪ Additionally, subtract any additional time outside the sprint for planned absences such as vacation time

  o Example: Eight-hour workdays with no planned away time for a two-week (10 day) sprint with five Development Team members

    ▪ It's calculated as:

      > 5 Development Team members x (8 hours in workday x [1 - 0.30 for time outside the sprint]) x 10 days in the sprint

      > = 5 x (8 x 0.70) x 10

      > = 5 x 5.60 x 10

      > = 280 project productive hours for sprint

5. If there are product release backlog items that have not yet been estimated in story points, that's taken care of now

   - For instructions, see **Section 4B: Estimation via Story Points**

6. If there are product release backlog items that have not yet been prioritized or need to be re-prioritized, the Product Owner takes care of that now

- For instructions, see **Section 4A: Product Release Backlog with User Stories and Prioritized Entries**

7. For each item in product release backlog, participants discuss the user stories/requirements including:

   - Acceptance criteria

   - Assumptions

   - Dependencies

   - Risks

   - Anything else requiring a conversation to get a good understanding of the item

8. If needed or helpful to understand the user stories/requirements, participants may also draw diagrams

9. The Development Team members commit to the entries which they think can be completed within the upcoming sprint

   - A simple technique:

     o The team asks, "Can we do this first product release backlog item in the sprint?"

       ▪ If the answer is yes, they select it and proceed to the next item and continue until the team believes that no more work can be done in the sprint

   - The Development Team considers story points and velocity:

       ▪ Each item in the backlog has an associated story point value

       ▪ If the team has worked together before, it has a historical velocity value

There's supplemental information in the following section of the appendix:

Velocity

       ▪ The team should consider their velocity when assessing backlog items can be done in the sprint

- Here's an example:

  - The sprint is two weeks, and the team's historical average velocity is 25 story points for a two-week sprint

  - The story points for the first five items in the product release backlog total 24 story points, so the Development Team is likely to say "yes" to those first five items

10. The Product Owner updates the product release backlog, designating the items committed for the upcoming sprint

Up next is **Section 6: Sprint Planning Part 2 (Sprint Backlog).**

It's recommended that **Sprint Planning Part 1** and **Sprint Planning Part 2** occur on the same day, with a break such as lunch between the two sessions.

Section 6: Sprint Planning Part 2 (Sprint Backlog)

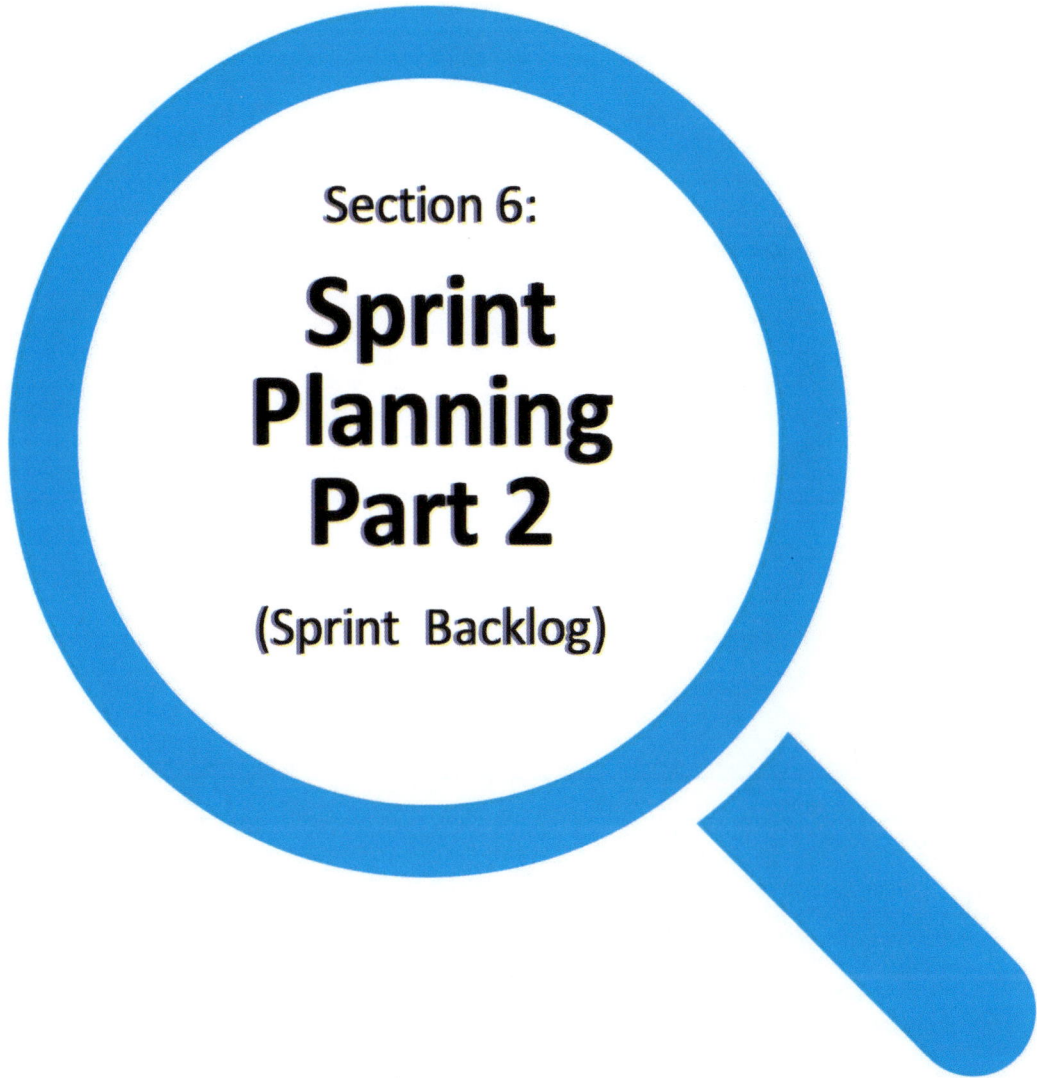

Section 6:

Sprint Planning Part 2

(Sprint Backlog)

Agile Scrum Cycle

Time-boxes Shown are Based on the Most Common Sprint Duration of Two Weeks
© Scott M. Graffius

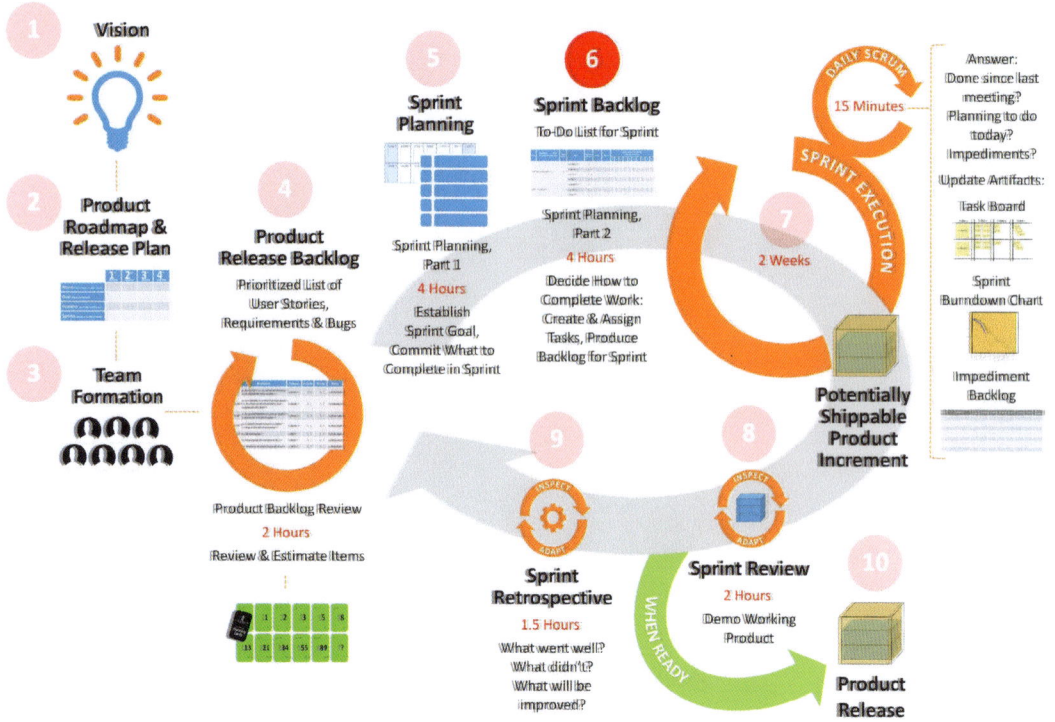

1 Vision

2 Product Roadmap & Release Plan

3 Team Formation

4 Product Release Backlog
Prioritized List of User Stories, Requirements & Bugs

Product Backlog Review
2 Hours
Review & Estimate Items

5 Sprint Planning
Sprint Planning, Part 1
4 Hours
Establish Sprint Goal, Commit What to Complete in Sprint

6 Sprint Backlog
To-Do List for Sprint
Sprint Planning, Part 2
4 Hours
Decide How to Complete Work: Create & Assign Tasks, Produce Backlog for Sprint

7 SPRINT EXECUTION
2 Weeks

DAILY SCRUM
15 Minutes
Answer: Done since last meeting? Planning to do today? Impediments?
Update Artifacts: Task Board
Sprint Burndown Chart
Impediment Backlog

Potentially Shippable Product Increment

8 INSPECT ADAPT
Sprint Review
2 Hours
Demo Working Product

9 INSPECT ADAPT
Sprint Retrospective
1.5 Hours
What went well? What didn't? What will be improved?

WHEN READY

10 Product Release

Purpose

During sprint planning, the Development Team pulls items from the top of the product release backlog to create a sprint backlog (part 1—the "what" meeting) and then decides how to accomplish those items during the next sprint (part 2—the "how" meeting). Part 1 was presented in the prior section. Part 2 is covered here.

This topic—**Section 6: Sprint Planning Part 2 (Sprint Backlog)**—is tagged with **6** in the illustration above.

Participants

- Scrum Master

- Product Owner

- Development Team

There's supplemental information in the following sections of the appendix:

| Product Owner | Scrum Master | Development Team | Self-Organization |

Frequency or sequence

After **Sprint Planning Part 1**

Time-box (not-to-exceed duration)

Time-boxed to two hours for each week of the sprint

Prerequisites or inputs

Completion of **Section 5: Sprint Planning Part 1**

Steps

1. The Development Team creates a sprint backlog containing the items selected during the Sprint Planning Part 1 meeting

    - A sample template is shown below

Sprint Backlog

ID #	Description (User Story, Requirement, Bug or Other Item)	Story Points	Task	Estimate (Hours)	Owner	Status	Hours of Work Remaining (by Day of Sprint)									
							1	2	3	4	5	6	7	8	9	10

- Notes on the fields follow:

    o ID#, description, and story points information come from the product release backlog

    o Task information is explained in step 2 of this section

    o Estimate (hours) are identified in step 3 of this section

    o Owner information is identified in step 4 of this section

    o Status information (such as To Do, Doing, Done) is covered in **Section 7: Sprint Execution**

    o Hours of work remaining for not-yet-done tasks are updated each workday of the sprint

- If you're new to Scrum, you may find it best to begin with the use of flip chart paper, a whiteboard or a simple spreadsheet

    o Specialized software tools—such as those listed in **Select Resources**—can always be used later

2. Starting with the first item, the Development Team identifies the tasks that need to be completed on it, and they update the sprint backlog

- Examples may include:

    o Meetings

    o Designs

    o Coding

    o Code review

    o Testing

    o Documentation

    o Or anything else that's required

3. The Development Team estimates the effort (in work hours) required for each task and updates the sprint backlog

- Suggestion:

- o If a task has effort greater than 8 hours, split the task into smaller ones

4. Development Team members sign themselves up for work and update the sprint backlog

- Work is never assigned by somebody else

There's supplemental information in the following section of the appendix:

Self-Organization

5. The Development Team repeats steps 2-4 for each item in the sprint backlog

6. If needed, the Development Team may make changes to the sprint backlog during this meeting (but not after the meeting)

- During the meeting:
    - o Compare the total estimated work hours for the sprint with the Development Team's capacity for the sprint
        - For the total estimated work hours for the sprint, see step 3 of this section
        - For capacity, see step 4 from **Section 5: Sprint Planning Part 1**
        - Here's an example:
            - The total estimated work hours for the sprint is 270
            - The Development Team's capacity for the sprint is 280 work hours
            - The team commits to the work because 270 hours is proximate to and within their capacity of 280
    - o If the Development Team believes that the sprint backlog contains too much work to be done during the sprint, they collaborate with the Product Owner to remove one or more items

o   If the Development Team believes that they can handle more work during the sprint, they collaborate with the Product Owner to move one or more of the most important remaining items from the product release backlog into the sprint backlog

Up next is **Section 7: Sprint Execution**.

Section 7: Sprint Execution

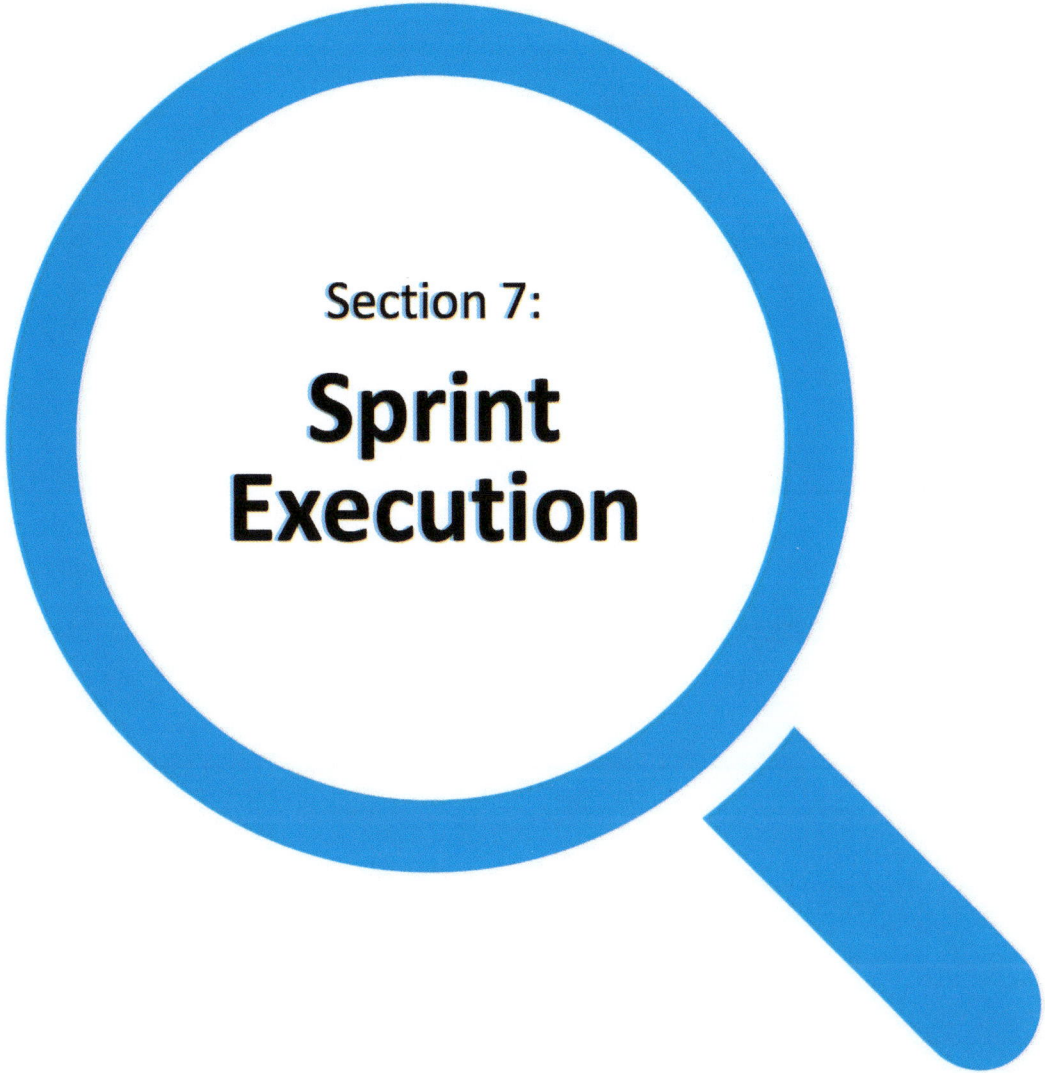

Section 7:

Sprint Execution

Agile Scrum Cycle

Time-boxes Shown are Based on the Most Common Sprint Duration of Two Weeks
© Scott M. Graffius

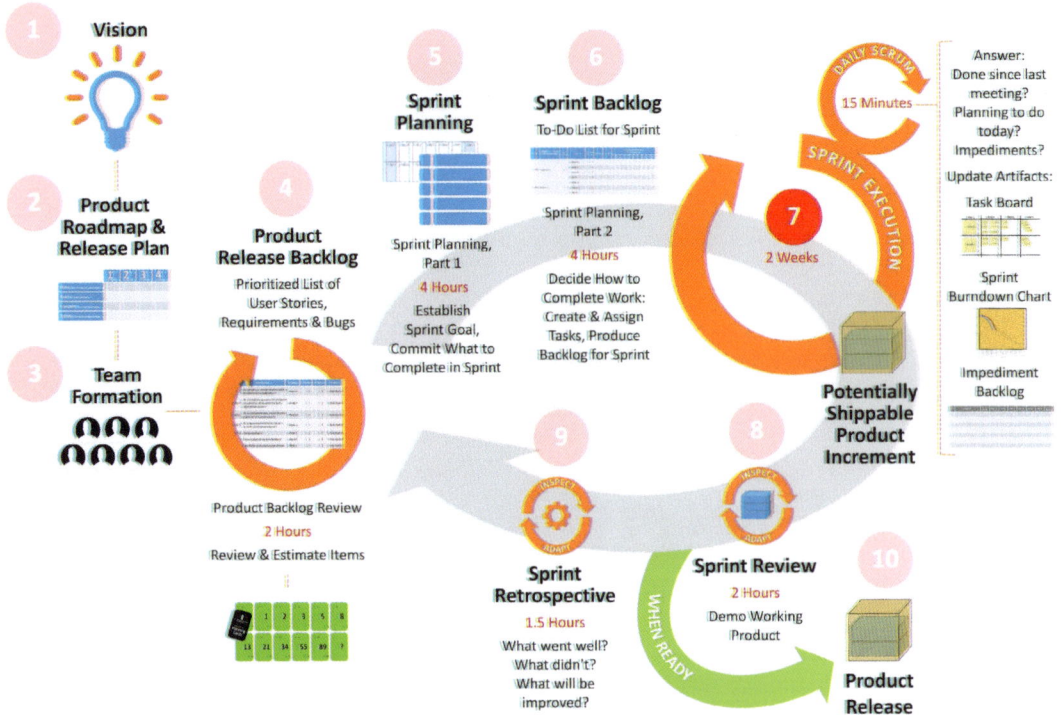

The Development Team works to complete the deliverables during Sprint Execution. What's needed varies by project. A software project—for example—may involve business analysis, technical designs, coding/pair programming with continuous integration, testing, user documentation, and other work.

The sprint backlog—covered in **Section 6: Sprint Planning Part 2 (Sprint Backlog)**—lists the work for the sprint.

Each workday during the sprint, there's a daily Scrum meeting, and the task board, sprint burndown chart, and impediment backlog are updated. A potentially shippable product increment is produced at the end of the sprint.

This topic—**Section 7: Sprint Execution**—is labeled with **7** in the illustration above.

Quick note

To make Sprint Execution easier to follow, there are five subsections:

- **7A: Task Board**

- **7B: Sprint Burndown Chart**

- **7C: Impediment Backlog**

- **7D: Daily Scrum Meeting**

- **7E: Potentially Shippable Product Increment**

Up next is **7A: Task Board**.

(This page is intentionally blank.)

7A: Task Board

Purpose

At the start of each sprint, the task board is set to reflect the work in the iteration. The task board is sometimes called the Scrum board.

Participants

- Development Team

- Scrum Master (facilitator)

> There's supplemental information in the following sections of the appendix:
>
> | Development Team | | Scrum Master |

Frequency or sequence

- Set-up occurs at the start of each sprint, as a follow-up to preparations covered during **Section 6: Sprint Planning Part 2 (Sprint Backlog)**

- Updated each workday

Time-box (not-to-exceed duration)

There's no time-box for this activity

Prerequisites or inputs

Section 6: Sprint Planning Part 2 (Sprint Backlog)

Steps

1. At the start of the sprint, the Development Team creates the task board based on the sprint backlog

    - Guidelines follow:

        o The task board depicts work in rows and columns

        o Rows include work items

        o Columns depict status (To Do, Doing, Done)

- Suggest keeping it simple, such as what's noted above

- Some organizations have variations, for example:

  - Different labels—like "In Progress" instead of "Doing"—may be used

  - Some add columns, such as having one for "Testing"

o Work is addressed from top (highest priority) to bottom

o Work migrates from left to right on the task board as it progresses

- This is explained further in **Section 7D: Daily Scrum Meeting**

o Task boards can be information that is hand-drawn on flip chart paper, sticky notes or index cards; alternatively, software tools can be used

- If you're new to Agile Scrum, you may find it best to begin with a simple hand-drawn task board

- Specialized software tools—such as those listed in **Select Resources**—can always be used later

o A task board of a project in progress will look something like this:

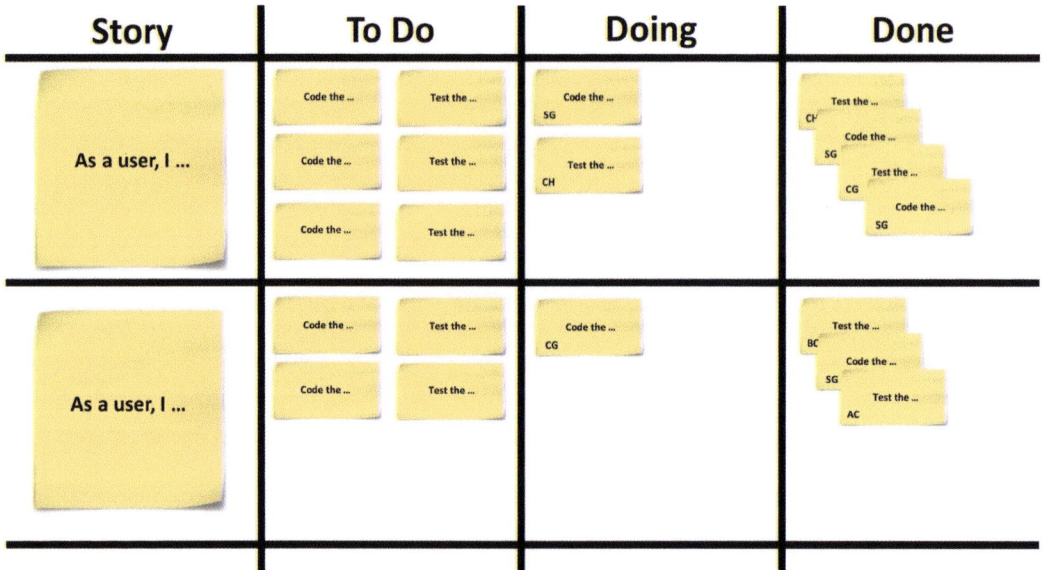

Story	To Do		Doing	Done
As a user, I ...	Code the ... Test the ... Code the ... Test the ... Code the ... Test the ...		Code the ... SG Test the ... CH	Test the ... CH Code the ... SG Test the ... CG Code the ... SG
As a user, I ...	Code the ... Test the ... Code the ... Test the ...		Code the ... CG	Test the ... BC Code the ... SG Test the ... AC

2. During sprint execution, the Development Team updates the task board

    o   This is covered in **7D: Daily Scrum Meeting**

There's supplemental information in the following section of the appendix:

Information Radiator

Up next is **7B: Sprint Burndown Chart**.

(This page is intentionally blank.)

7B: Sprint Burndown Chart

Purpose

> Scrum teams use the sprint burndown chart to track and communicate progress during the sprint.

Participants

- Development Team

- Scrum Master (facilitator)

> There's supplemental information in the following sections of the appendix:
>
> | Development Team | | Scrum Master |

Frequency or sequence

- Set-up at the start of the sprint

- Updated each sprint workday

  o Typically, before the daily Scrum meeting

Time-box (not-to-exceed duration)

> There's no time-box for this activity

Prerequisites or inputs

> Completion of **7A: Task Board**

Steps

1. At the start of the sprint, the Development Team creates the sprint burndown chart

   - Guidelines follow

     o The sprint burndown chart involves:

       ▪ Horizontal x-axis displays working days

         — Such as sprint day 1 through day 10

- Vertical y-axis displays remaining work

    - In story points (this is most common), or

    - In hours

- Ideal velocity is included as a guideline

There's supplemental information in the following section of the appendix:

Velocity

- Actual progress-to-date

o These charts can be hand-drawn on flip chart paper or whiteboards, or software tools can be used

    - If you're new to Agile Scrum, you may find it best to begin with the use of flip chart paper or a whiteboard

    - Specialized software tools—such as those listed in **Select Resources**—can always be used later

o An example of a sprint burndown chart is shown next

Sprint Burndown Chart

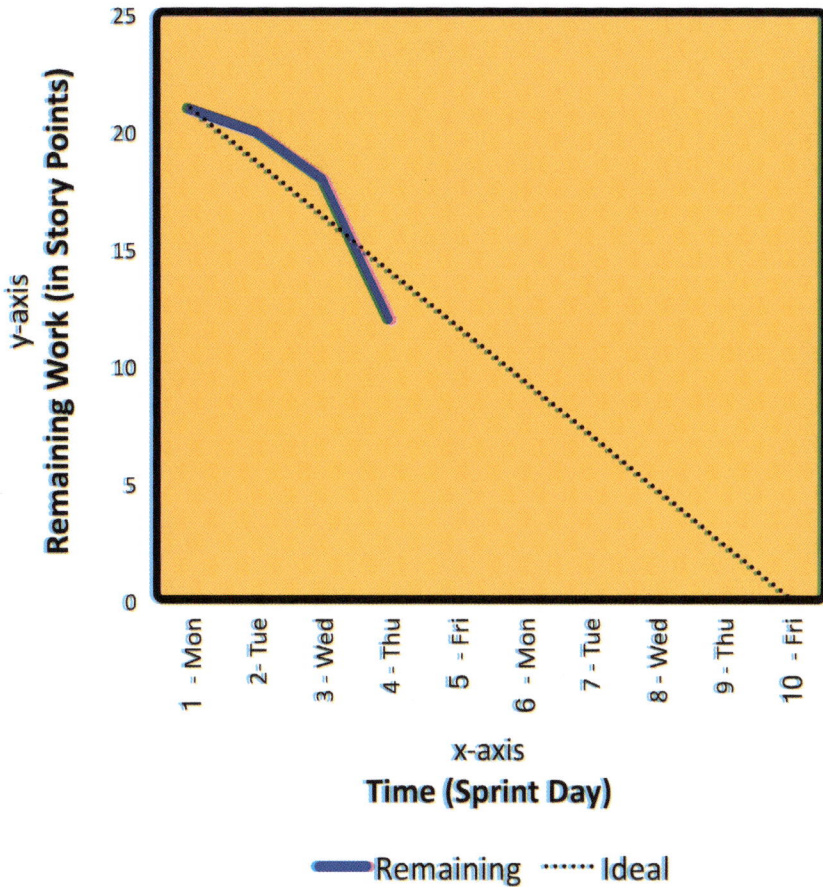

During the sprint, the Development Team updates the sprint burndown chart

- Typically, it's updated at the end of each workday

There's supplemental information in the following section of the appendix:

Information Radiator

Some organizations also have a product release burndown chart, updated by the Product Owner at the end of each sprint.

There's supplemental information in the following sections of the appendix:

Product Release Burndown Chart   Metrics

Up next is 7C: Impediment Backlog.

7C: Impediment Backlog

Purpose

The impediment backlog is a list of things that are preventing the team from progressing or improving. Impediments may also be known as blockers, issues or problems.

Here are a few examples of potential impediments:

- Product vision is unclear

- Sprint goal is unclear

- Missing or inadequate tools

- Product Owner is not available to answer questions

- The Development Team is too small or too big

    o There should be between three and nine (formerly between five and nine) people in the Development Team

- Artifacts—such as the task board, sprint burndown chart, and impediment backlog—are not clearly posted or easily accessible

- Conflicts

- Staff availability not meeting needs

- Insufficient expertise or knowledge

The Scrum team is responsible for raising impediments. The Scrum Master is responsible for helping to resolve the problems.

Participants

- Scrum Master (who searches for impediments and collaborates with others to resolve problems)

- Development Team

There's supplemental information in the following sections of the appendix:

Scrum Master   Development Team

Frequency or sequence

- Set up at the start of the sprint

- Updated each sprint workday

  o Typically, after the daily Scrum meeting

  o It may be updated more often such as also around the end of the workday

Time-box (not-to-exceed duration)

There's no time-box for this activity

Prerequisites or inputs

Completion of **Section 6: Sprint Planning Part 2 (Sprint Backlog)**

Steps

1. At the start of the sprint, the Scrum Master creates the impediment backlog

- Guidelines follow

  o The impediment backlog can be hand-drawn on flip chart paper or whiteboards, or software tools can be used

    ▪ If you're new to Agile Scrum, you may find it best to begin with the use of flip chart paper, a whiteboard or a simple spreadsheet

    ▪ Specialized software tools—such as those listed in **Select Resources**—can always be used later

  o Entries typically include:

    ▪ ID number

    ▪ Description

    ▪ The date the impediment was raised

    ▪ Who raised it

    ▪ Priority

- It should always be sorted, with open high-priority items shown first

- Actions

  - What's being done to address the problem

- Owner

  - It's the Scrum Master unless ownership is adopted by another individual

- Notes

  - Additional notes, if any

- Status

  - Open or

  - Closed

- Date resolved

- An impediment backlog sample template is shown next

Impediment Backlog

ID #	Description	Date Raised	Raised by	Priority	Actions	Owner	Notes	Status	Date Resolved

2. During sprint execution, the Scrum Master updates the impediment backlog each workday

- Typically, the impediment backlog is refreshed shortly after the daily Scrum meeting

  o This is discussed further in the next subsection

- It may be updated more often, such as also around the end of the workday

- Impediments can be managed via time limit or quantity

  o To limit time, set a time-box (24 hours, for example)

    ▪ Either the impediment is resolved within the time period, or it is closed as "trashed"

  o To limit quantity, set a maximum (three, for example) for open impediments

There's supplemental information in the following section of the appendix:

Information Radiator

Up next is **7D: Daily Scrum Meeting**.

7D: Daily Scrum Meeting

Purpose

The daily Scrum is a time for the self-organizing Development Team to share status, plans, and any impediments. The daily Scrum is sometimes referred to as the daily stand-up meeting. This meeting is a Scrum event (formerly called a ceremony).

There's supplemental information in the following section of the appendix:

Self-Organization

Participants

- Development Team

- Scrum Master

- Product Owner (optional, observer)

There's supplemental information in the following sections of the appendix:

Development Team   Scrum Master   Product Owner

Frequency or sequence

- Each sprint day

    o The daily Scrum is essential to the ongoing success of any sprint—it keeps the Development Team focused on deliverables and reinforces commitments to their fellow team members

    o The daily Scrum should not be canceled, even if only a few team members can attend on a given day

- Same time each workday

    o Typically, in the morning

Time-box (not-to-exceed duration)

Time-boxed to 15 minutes

o   To support keeping it brief, it's recommended that everyone remain standing for the meeting

Prerequisites or inputs

- Completion of **Section 6: Sprint Planning Part 2 (Sprint Backlog)**

- Completion of **7A: Task Board**

- Completion of **7B: Sprint Burndown Chart**

- Completion of **7C: Impediment Backlog**

Steps

1. The Development Team and Scrum Master meet at a location where the task board, sprint burndown chart and impediment backlog are visible

2. A Development Team member explains what he/she worked on since the last daily Scrum meeting, and he/she updates the task board

    - As appropriate, he/she updates tasks with status changes:

        o   Move tasks from "To Do" to "Doing"

        o   Move tasks from "To Do" to "Done" (when that leap is applicable) or

        o   Move tasks from "Doing" to "Done"

    - For any tasks with the status of "Doing," the estimated remaining work hours are noted

3. Next, the same Development Team member explains what he/she will work on that day

    - If the task already had the status of "Doing"

        o   The status remains "Doing"

        o   The Development Team member notes the estimated remaining work hours for the task

- If the task to be worked on had the status of "To Do" before

  o The status is updated to "Doing"

  o The Development Team member notes the estimated work hours for the task

4. Next, the same Development Team member reports any impediments

5. The Scrum Master records impediments in the impediment backlog

   - If discussion is required, it takes place immediately after the daily Scrum

     o (The Scrum Master helps resolve impediments after the daily Scrum)

6. Steps 2-4 are repeated for the other members of the Development Team, and the Scrum Master repeats step 5 when impediments are raised

Up next is **Section 7E: Potentially Shippable Product Increment.**

(This page is intentionally blank.)

7E: Potentially Shippable Product Increment

Scrum requires teams to build an increment of functionality during every sprint, and the increment must be potentially shippable because the Product Owner might decide to release it at the end of the sprint.

- The product increment is the sum of all backlog items completed during the current sprint

- Potentially shippable is defined by a state of confidence or readiness, and shipping is a business decision

    o Shipping may or may not occur at the end of the sprint

    o New functionality may be accumulated via multiple sprints before being shipped

---

Minimum Viable Product Approach

The product increment may or may not be marketable. However, a Minimum Viable Product (MVP) approach is sometimes used to help test marketable ideas.

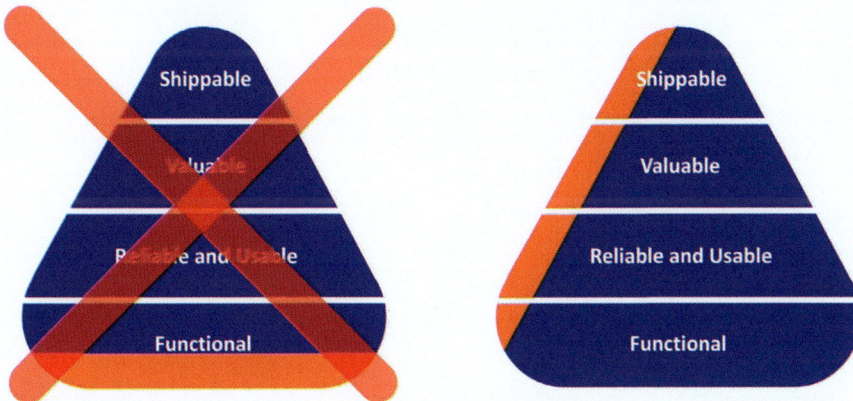

The MVP has just those features (functional, reliable and usable) considered sufficient for it to be of value to customers and allow for it to be shipped or sold to early adopters. Customer feedback will inform future development of the product.

The new functionality is demonstrated to the Product Owner and stakeholders during the sprint review (covered in the next section), and deployment is discussed further in **Section 10: Product Release**.

Up next is **Section 8: Sprint Review**.

Section 8: Sprint Review

Section 8:

Sprint Review

Agile Scrum Cycle

Time-boxes Shown are Based on the Most Common Sprint Duration of Two Weeks
© Scott M. Graffius

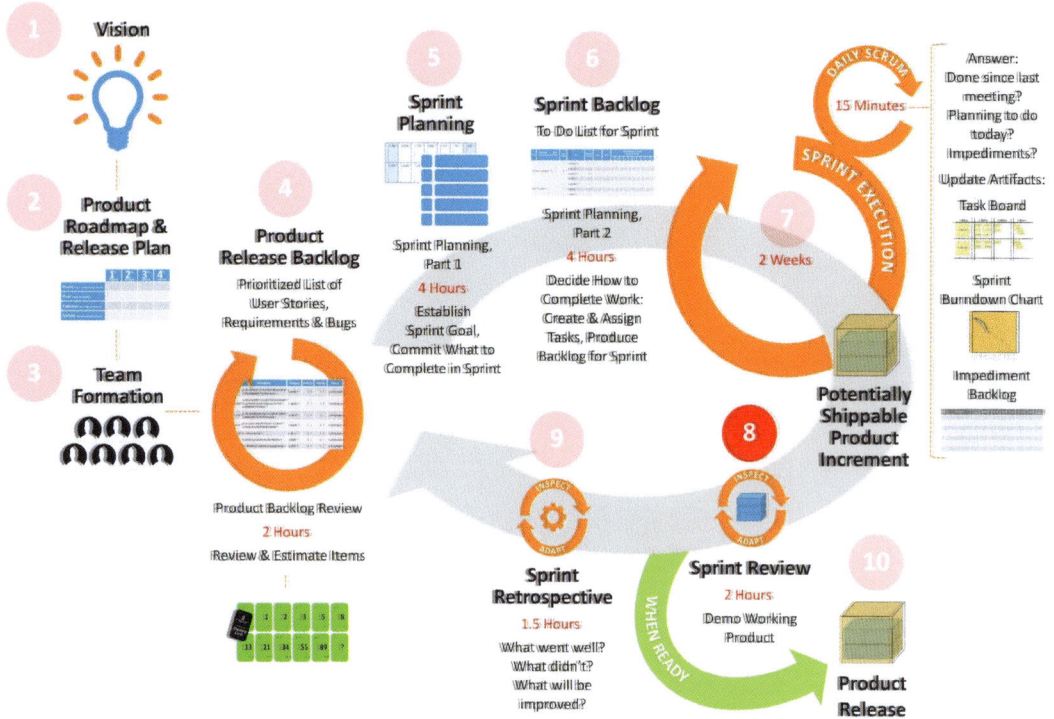

Purpose

A sprint review is held at the end of the sprint to go over outcomes, provide live demonstrations of "done" functionality and determine next steps. This meeting is a Scrum event (formerly called a ceremony).

This topic—**Section 8: Sprint Review**—is tagged with **8** in the illustration above.

Participants

- Product Owner

- Scrum Master

- Development Team

- Stakeholders

There's supplemental information in the following sections of the appendix:

| Product Owner | | Scrum Master | | Development Team | | Stakeholders |

| Inspect and Adapt |

Frequency or sequence

After completion of sprint execution

Time-box (not-to-exceed duration)

Time-boxed to one hour for each week in the sprint

Prerequisites or inputs

Completion of **Section 7: Sprint Execution**

Steps

1. The Product Owner welcomes attendees to the sprint review meeting and provides an overview of the agenda (such as these 11 points)

2. The Product Owner reminds attendees of the sprint goal

   - Recommend prominently displaying the sprint goal in the meeting room

3. The Development Team lists the work that was committed to the sprint

4. The Development Team lists the work that was completed during the sprint

5. The Development Team lists the work that wasn't completed during the sprint

6. For each story/feature, the Development Team demonstrates the "done" working functionality and answers questions about the increment

   - Only "done" working functionality is demonstrated

7. Stakeholders may use or interact with the "done" working functionality

8. Stakeholders provide feedback about the "done" working functionality

9. The entire group reviews the product release backlog and collaborates on what to do next

10. The Product Owner incorporates feedback into the product release backlog

    - This typically involves

        o Adding new items to the product release backlog, and/or

        o Re-prioritizing existing items

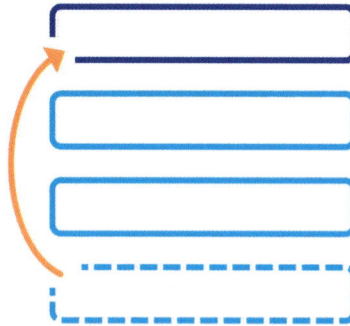

11. The Scrum Master incorporates feedback related to blockers, issues or problems into the impediment backlog

    - After the meeting, the Scrum Master helps to resolve the impediments

Up next is **Section 9: Sprint Retrospective.**

Section 9: Sprint Retrospective

Section 9:

Sprint Retrospective

Agile Scrum Cycle

Time-boxes Shown are Based on the Most Common Sprint Duration of Two Weeks
© Scott M. Graffius

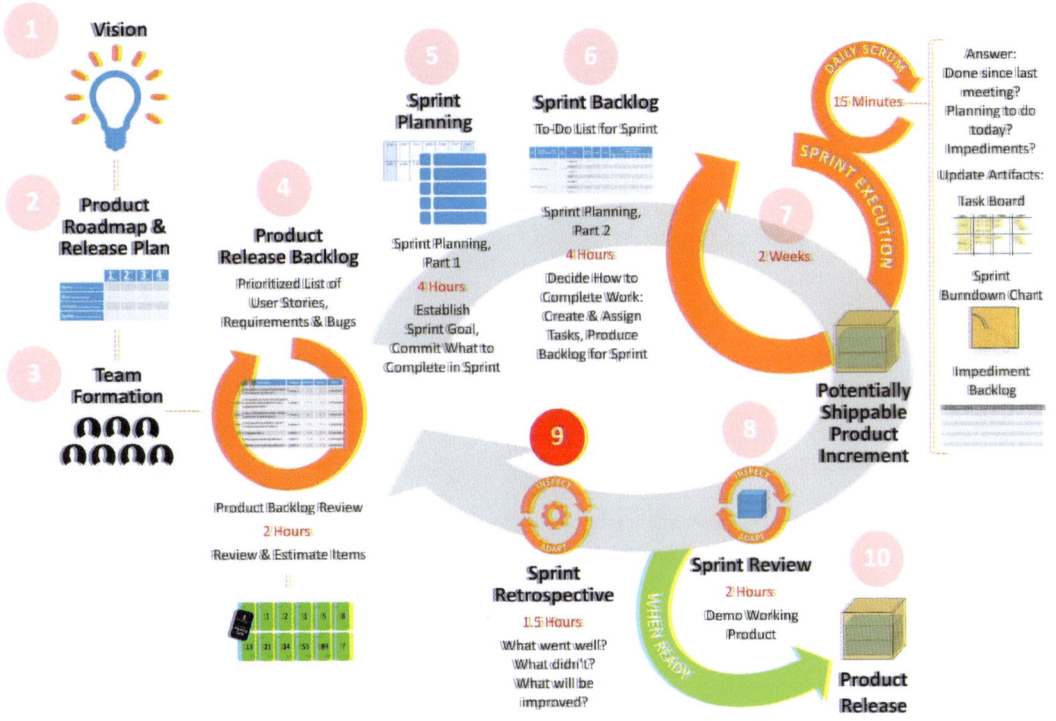

1 Vision

2 Product Roadmap & Release Plan

3 Team Formation

4 Product Release Backlog

Prioritized List of User Stories, Requirements & Bugs

Product Backlog Review
2 Hours
Review & Estimate Items

5 Sprint Planning

Sprint Planning, Part 1
4 Hours
Establish Sprint Goal, Commit What to Complete in Sprint

6 Sprint Backlog
To-Do List for Sprint

Sprint Planning, Part 2
4 Hours
Decide How to Complete Work: Create & Assign Tasks, Produce Backlog for Sprint

7 2 Weeks

DAILY SCRUM
15 Minutes
SPRINT EXECUTION

Answer:
Done since last meeting?
Planning to do today?
Impediments?
Update Artifacts:
Task Board
Sprint Burndown Chart
Impediment Backlog

Potentially Shippable Product Increment

9 Sprint Retrospective
1.5 Hours
What went well?
What didn't?
What will be improved?

8 Sprint Review
2 Hours
Demo Working Product

WHEN READY

10 Product Release

Purpose

At the sprint retrospective meeting, the Scrum Master and the Development Team inspect the performance of the preceding sprint and identify what went well, what didn't go well, and improvements to be implemented during the next sprint. This meeting is a Scrum event (formerly called a ceremony).

This topic—**Section 9: Sprint Retrospective**—is labeled with **9** in the illustration above.

Participants

- Scrum Master

- Development Team

> There's supplemental information in the following sections of the appendix:
>
> [Scrum Master]   [Development Team]   [Inspect and Adapt]

This meeting may be limited to only the Scrum Master and the Development Team

Frequency or sequence

After the sprint review meeting

Time-box (not-to-exceed duration)

Time-boxed to 0.75 hour for each week of the sprint

Prerequisites or inputs

Completion of **Section 8: Sprint Review**

Steps

Part 1: Identifying What Went Well

Here's a common approach:

1. The Scrum Master distributes sticky notes and pens to Development Team members

2. The Scrum Master asks each Development Team member to write the top one, two or three things that went well during the preceding sprint on sticky notes

   - One item per sticky note

3. Each Development Team member writes down the top one, two or three things

   - It's best for the team to formulate ideas without any additional guidance

     o If they are unable to arrive at any items, the Scrum Master may relate examples of topics from prior sprint retrospectives for reflection

       ▪ Possible examples follow:

         − Communications with those outside the core team

         − Communications within the core team

         − Definition of Done

          – External distractions (free from them or how they were handled well)

          – Input from stakeholders

          – Performance shown on the sprint burndown chart

          – Sprint backlog items completed vs. not completed

          – Team composition

          – Team skill sets

          – Tools

          – Velocity

4. The Development Team places the sticky notes on a flip chart, whiteboard or wall

5. The Development Team groups the sticky notes into categories/themes

6. Participants reflect on and discuss their successes

7. Participants commit to continue their successful practices

Part 2: Identifying What Didn't Go Well and Improvements to be Implemented

Here's a common method:

1. The Scrum Master distributes sticky notes and pens to Development Team members

2. The Scrum Master asks each Development Team member to write the top one, two or three things that didn't go well on sticky notes

      • One item per sticky note

3. Each Development Team member writes down the top one, two or three things

      • It's best for the team to formulate ideas without any additional guidance

          o But if they are unable to arrive at any items, the Scrum Master may relate examples of topics from prior sprint retrospectives for reflection

              ■ Possible examples follow:

                  – Communications with those outside the core team

- Communications within the core team

- Definition of Done

- External distractions

- Input from stakeholders

- Performance shown on the sprint burndown chart

- Sprint backlog items completed vs. not completed

- Team composition

- Team skill sets

- Tools

- Velocity

4. The Development Team places the sticky notes on a flip chart, whiteboard or wall

5. The Development Team groups the sticky notes into categories/themes

6. Of the things that didn't go well, participants brainstorm on potential solutions to be incorporated in the next sprint (the team may want to consider how they can make a bigger impact or how they can achieve greater ease of work)

7. Participants vote to determine the top one or two items for adaptation and improvement

8. Participants commit to making the changes

9. The Scrum Master records the information

More Options for Sprint Retrospective Meetings

Alternative approaches may be taken at the retrospectives. Here are a few examples:

- Awards Ceremony: The Development Team establishes a few categories—in an awards program motif—such as Special Achievement Award, Best User Story Award, etc. For each one, they determine the placements, and they discuss the rationale in depth. They then decide what will be improved upon in the next sprint.

- Envisioning the Future: The Development Team is asked to imagine that their next sprint went perfectly. They are asked to describe it including what was done differently to achieve the excellent performance. They then decide what will be changed in their next sprint to support improvement.

- Start/Stop/Continue: The Development Team identifies what they should start doing, stop doing, and continue doing. They then decide what will be improved upon in their next sprint.

- Red/Green: The Development Team establishes a few categories such as Delivering Value, Speed, etc. For each one, they arrive at an understanding of what represents green (favorable) and red (not favorable). Here's an example for Speed: green represents "we get stuff done quickly" while red means "we never seem to get anything done." The team votes on each category and discusses their views. They then decide what will be improved upon in the next sprint.

In addition to the sprint retrospective meeting, some organizations employ surveys on satisfaction and other metrics and track trends over time.

There's supplemental information in the following section of the appendix:

Metrics

Up next is **Section 10: Product Release**.

Section 10: Product Release

Section 10:

Product Release

Agile Scrum Cycle

Time-boxes Shown are Based on the Most Common Sprint Duration of Two Weeks
© Scott M. Graffius

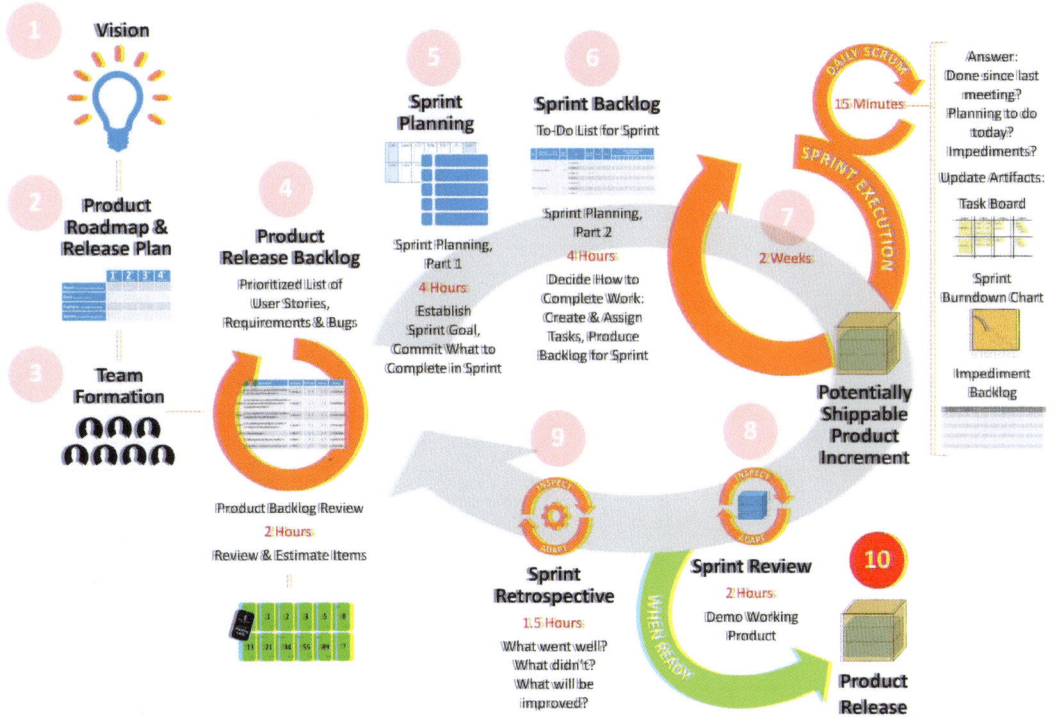

This topic—**Section 10: Product Release**—is annotated with [10] in the illustration above.

Here's a quick recap of the prior nine sections:

1. The product vision statement was created

2. The combination product roadmap and release plan was established

3. The Scrum team was formed

4. The product release backlog was developed, and items were estimated in story points and prioritized

5. The sprint goal was established, and the Development Team committed to what would be completed in the sprint

6. The Development Team determined how to complete the work, and the sprint backlog was produced

7. Sprint execution encompassed the work of the Development Team and supporting activities and artifacts—the daily Scrum, task board, sprint burndown chart, and impediment backlog; most importantly, a potentially shippable product increment was produced

8. The sprint review was the venue for live demonstrations of the done functionality

9. The sprint retrospective facilitated the identification of one or more action items supporting continuous improvement

Product Release

Scrum emphasizes a working product at the end of each sprint, and increments may or may not be shipped at the conclusion of each sprint. The table below outlines product release scenarios.

Product Release After One or More Sprints	
Release at the End of a Sprint	**Release After Multiple Sprints**

• Each sprint is between one and four weeks in length (the most common duration is two weeks)
 	
• Each sprint produces a potentially shippable product increment
 	
• The product may or may not be shipped at the end of the sprint
 	
• Shipping is a business decision
	• Example of release after multiple sprints:
	  - Duation of each sprint is two weeks
	  - Involved three sprints
	  - Product release occurred after six weeks (sprint duration of two weeks x three sprints)
	
	• What's shipped to customers/users is the sum of all the product release backlog items completed during the sprints, and it contains the value of the increments of all previous sprints

As covered in sections 8 and 9, sprint review and sprint retrospective meetings occur at the end of each sprint. After the final release, some organizations also have a review and retrospective of the full project.

That's it for **Section 10: Product Release**. And that concludes the instructions on how to successfully employ the Agile Scrum framework. Congratulations!

Additional Content

There is additional information—including examples, illustrations, tips, and more—in the subsequent sections of this guide.

You can skip them if you'd like, but the extra content is informative, concise (items in the appendix are about a page per topic) and there for your reference whenever you need it.

Glossary of Terminology

Acceptance criteria: Details that indicate the scope of a user story and help the team determine "done"-ness.

Agile: An iterative and incremental development method. It promotes evolutionary development and is designed to support rapid and flexible responses to changing requirements.

Agile Manifesto: A formal proclamation of four key values and twelve principles to guide an iterative and people-centric approach to development.

Agile principles: The twelve principles that underpin the Agile Manifesto. See **Agile Manifesto**.

Artifacts: Documents or visual depictions of work items, progress, features, code base, etc.

Automated integration testing: Where individual software modules are automatically combined and tested as a group.

Backlog: See **Product release backlog** and **Sprint backlog**.

Blockers: See **Impediment**.

Burndown: See **Sprint burndown chart** and **Product release burndown chart**.

Burn rate: The rate at which hours allocated to a project are being used.

Cadence: A regular, predictable rhythm. Sprints of consistent duration (two weeks, for example) establish a cadence for development effort.

Cancellation: See **Termination**.

Ceremonies: See **Events**.

Chart: See **Sprint burndown chart** and **Product release burndown chart**.

Continuous integration: Where individual software modules are combined and tested with existing software as soon as they are produced.

Core roles: There are three core roles—Product Owner, Scrum Master, and Development Team. The core roles collectively constitute the Scrum team.

Cross-functional team: A group of people with different expertise working towards a common goal.

Daily Scrum: Each day during a sprint, the team holds a daily Scrum meeting. It lasts no more than fifteen minutes, and it provides a quick update on progress. Each member of the Development Team reports on tasks finished the prior workday, work to be done today, and any impediments blocking progress. This event is sometimes referred to as the daily stand-up or the daily sync.

Daily Stand-up: See **Daily Scrum**.

Definition of Done (DoD): The exit-criteria to determine whether a product backlog item is complete. It provides precision about when work is complete. The DoD typically involves the following: Code complete, unit tests written and executed, integration tested, performance tested, documented, and accepted by Product Owner. The DoD may vary from one Scrum team to another, but it must be consistent within a team.

Demo: See **Sprint review**.

Development Team: A cross-functional group of people responsible for delivering potentially shippable increments of a product at the end of every sprint. The team is comprised of between three and nine (a previous guideline was between five and nine) people—developers, testers, business analysts, etc. Their responsibilities include being self-organizing, negotiating commitments with the Product Owner, collaborating with anyone necessary to get the job done, delivering the product, alerting the Scrum Master of any impediments, presenting the product at the sprint review (demo), and inspecting and adapting the process.

DoD: See **Definition of Done**.

Done: See **Definition of Done**.

Estimated work remaining: The hours that a team member estimates that remains to be worked on for a task. This estimate is updated each day that the task is worked on.

Events: There are four key events (meetings)—Sprint planning, Daily Scrum, Sprint review, and Sprint retrospective. Events were previously known as ceremonies. Also see: **Sprint planning**, **Daily Scrum**, **Sprint review**, and **Sprint retrospective**.

Fibonacci sequence: Attributed to Italian mathematician Leonardo Pisano, the Fibonacci sequence is a series of numbers where a value is found by adding up the two numbers before it. Starting with 0 and 1, the sequence goes 0, 1, 1, 2, 3, 5, 8, 13, 21, 34, etc. In Agile Scrum, story points are used to rate the difficulty to implement a story (work). Most Scrum teams use the Fibonacci sequence or a variation as an estimating technique.

Functionality: The behaviors that a computer system is designed to achieve.

Impediment: Issues or anything else that prevents a team member from performing work as efficiently as possible. Impediments may also be known as blockers, issues or problems.

Increment: Potentially shippable completed work that is the outcome of a Sprint.

Information radiator: A large graphical representation of key project information, updated regularly. The information is displayed near the team's workspace, accessible to stakeholders. The information may be presented on flip chart paper or a whiteboard or displayed on a monitor. Information radiators are sometimes called wallboards or electronic wallboards.

INVEST: Introduced by Bill Wake, INVEST is an acronym for a model for developing a well-written user story. INVEST stands for Independent, Negotiable, Valuable, Estimable, Small, and Testable. Also see: **User story**.

Issue: See **Impediment**.

Iteration: A short time period in which a team is focused on delivering an increment of a product that is useable. Also see: **Sprint**.

Meetings: See **Events**.

Minimum viable product: The MVP has just those features (functional, reliable and usable) considered sufficient for it to be of value to customers, and allow for it to be shipped or sold to early adopters. Customer feedback will inform future development of the product.

MoSCoW: MoSCoW is a popular acronym that represents a method of ranking stories (requirements, new features, bug fixes, etc.). With MoSCoW, each item is sorted into one of four categories: Must have, Should have, Could have, or Won't have. Also see: **User story**.

Pair programming: An agile software development technique in which two programmers collaborate on the same task on a single computer. The person who controls the mouse and keyboard is called the "driver." The other person, who sits beside the driver and helps ensure that the solution is implemented in an effective and efficient manner, is called the "navigator." The members of the pair can switch with other members of the team to take on different tasks. The key objective is to produce high-quality code. Benefits typically include no separate code reviews required, better code quality, effective communication, better coding practice adherence, a more effective inclusion of new team members, and greater knowledge sharing among team members.

Potentially shippable product: An increment of work that is complete per the Definition of Done and is capable of being released.

Problem: See **Impediment**.

Product backlog grooming: See **Product release backlog refinement**.

Product goal: See **Product vision statement**.

Product increment: The increment or potentially shippable increment (PSI) is the sum of all the product backlog items completed during a sprint and all previous sprints. At the end of a sprint, the increment must be complete, according to the Scrum team's Definition of Done, and in a usable condition regardless of whether the Product Owner decides to release it.

Product Owner: The person responsible for maintaining the product backlog by representing the interests of the stakeholders and ensuring the value of the work done by the Development Team. More specifically, the Product Owner is responsible for representing the voice of customers, communicating with stakeholders to ensure their interests are represented, managing stakeholder expectations, establishing and achieving the product vision, defining releases, defining sprint goals, managing the return on investment, creating and maintaining the product backlog, authoring and prioritizing user stories based on business value, outlining acceptance criteria, attending the sprint reviews and planning sessions, and continuously reprioritizing the product backlog.

Product release backlog: A list of functional and non-functional requirements, usually expressed as user stories. Entries are prioritized based on business value.

Product release backlog refinement: An ongoing process of adding detail and estimates, as well as re-ordering the backlog items. It is sometimes referred to as **Product backlog grooming**.

Product release burndown chart: This depicts the points of all user stories from the product backlog. It shows the story points for completed work in each sprint, illustrating the completion of requirements over time. Also see: **Sprint burndown chart** and **Information radiator**.

Product vision statement: The product vision is the overall objective. It's created by the Product Owner, and it describes in about two sentences the product's purpose, who it is for, how it will create value, and the benefits. The vision often encompasses multiple releases. The product vision is sometimes called the **Product goal**.

Refactoring: The process of improving code—by clean-up, simplification, etc.—to make it easier to maintain and expand in the future. Refactoring is a necessary step to keep the cost of changes low.

Release: The transition of the final product into routine use by the end user.

Retrospective: See **Sprint Retrospective**.

Scrum: The most popular agile development and delivery framework. It encompasses a powerful set of principles and practices that help teams deliver products in short cycles, which promotes speedy feedback, rapid adaptation to change, faster delivery time, and continuous improvement.

Scrum components: Scrum's roles, events, artifacts, and the rules that bind them together.

Scrum Master: The individual who ensures the team adheres to Scrum practices, values, and rules. His/her responsibilities include serving as single point of contact for the project, facilitating and enforcing the Scrum process, coaching the team on Scrum values and practices, facilitating the daily Scrum (stand-up meeting), ensuring the team is fully functional and productive, helping the team remove impediments, capturing data, keeping Scrum artifacts (charts, etc.) current and visible, enforcing time-boxes, promoting improved engineering practices, shielding the team from external interferences and distractions, and conducting the sprint retrospective at the end of a sprint.

Scrum team: The Scrum team is made up of the Product Owner, Scrum Master, and Development Team. It totals between five and eleven (between seven and eleven in a prior guideline) people.

Scrum values: A set of fundamental values and qualities underpinning the Scrum framework: focus, courage, openness, commitment, and respect.

Self-organizing: The principle that teams autonomously organize their work. Self-organization happens within boundaries and against given goals. Self-organizing teams determine how to best accomplish their work, rather than being directed by others outside the team.

Show and tell: See **Sprint review**.

Spike: A time-boxed period used to research a concept or create a prototype. Spikes typically occur between sprints. Unlike sprints, spikes may or may not deliver shippable and valuable functionality.

Sprint: A period of one, two, three or four weeks (two is most common) during which the team will work on a set of backlog items that were committed to be completed.

Sprint backlog: A prioritized list of tasks to be completed during the sprint. Also see: **MoSCoW**.

Sprint burndown chart: Shows the work remaining in the sprint backlog. It is refreshed before the next daily Scrum meeting. Also see: **Product release burndown chart** and **Information radiator**.

Sprint goal: The purpose of a Sprint. It's often expressed as a proposed solution to a business problem.

Sprint planning: A time-boxed event (meeting), set to run two hours for each week of the sprint. It occurs before the sprint, and there are two parts to the event. During the first half, the sprint team (Product Owner, Scrum Master, Development Team) agree on what product backlog items to consider for the sprint. During the second half, the Development Team decomposes the work required to deliver the backlog items, resulting in the sprint backlog.

Sprint retrospective: At the end of a sprint, the team holds two meetings—the sprint review and the sprint retrospective. At the sprint retrospective, the team reflects on the past sprint, and identifies and prioritizes improvements for the next sprint. The event is time-boxed to run 45 minutes for each week of the sprint and is facilitated by the Scrum Master. Also see: **Sprint review**.

Sprint review: At the end of a sprint, the team holds two events (meetings)—the sprint review and the sprint retrospective. At the sprint review, the team reviews the work that was completed and the planned work that was not completed, and demonstrates the completed work. The event is time-boxed to run one hour for each week of the sprint. Also see: **Sprint retrospective**.

Sprint tasks: Work items added to the sprint backlog at the beginning of a sprint and broken down into hours. It is recommended that each task not exceed six hours (or one work day).

Stakeholder: Someone with an interest in the outcome of a project, either because he or she has funded it, will use it or will be affected by it.

Stand-up meeting: See **Daily Scrum**.

Story points: Used to rate the difficulty (related to complexity, unknowns, etc.) to implement a story. The Fibonacci sequence (0, 1, 2, 3, 5, 8, 13, 21, ...) and t-shirt sizes (small, medium, large, extra-large) are common scales. Also see: **User story** and **Fibonacci sequence**.

Task board: A notice board that shows the progress of each task. The most basic states for status are: "To Do", "Doing," and "Done." The task board can be physical or shared via a software solution.

Tasks: See **Sprint tasks**.

Team: See **Scrum team** and **Development Team**.

Team capacity, estimated: This is calculated as the number of sprint Development Team members, multiplied by the number of project productive hours per workday, multiplied by the number of work days in the sprint. Project productive hours should exclude time outside the sprint, such as company meetings, trainings, vacation time, etc.

Technical debt: The cumulative total of poor design and coding. It may have one or more causes such as time pressures, overly complex technical design, lack of alignment to standards, suboptimal code, delayed refactoring, insufficient testing or inadequate documentation. The consequence of technical debt is that more time is needed later on in the project to resolve issues. It can be avoided or minimized by not taking shortcuts, using simple designs, and refactoring continuously. When there's technical debt, the team should make the items visible by registering entries in the product release backlog, where the items will be evaluated and prioritized for resolution.

Termination, Abnormal: The Product Owner can cancel a sprint if necessary. Changes in external market conditions may be a reason, for example. If a sprint is abnormally terminated, the next step is to conduct a new sprint planning session, where the reason for the termination is reviewed.

Testing: In Agile Scrum, the approach includes testing performed early and often, and close cooperation with developers and customers/users. It often involves unit testing, application program interface and service testing, acceptance testing, system testing, regression testing (to detect any side effects from changes), and user acceptance testing (UAT). Tests should be automated as much as possible.

Time-box: Setting a duration for an activity and not exceeding it. For example, a daily Scrum meeting is time-boxed to 15 minutes and ends no later than that time.

User story: Each user story represents a portion of business value that a team can deliver in an iteration. They act as requirements and are written in plain language. The format of a user story is: "As a <Role>, I want <goal> so that I can <reason>." Example: "As a customer, I want shopping cart functionality so that I can buy items online." User stories are captured in the product release backlog. Also see: **INVEST**, **Story points** and **Fibonacci sequence**.

Velocity, Actual: The sum of the team's delivery of completed work from the product backlog. It is usually measured in story points. Example: A team delivers story "A," which had 4 points, story "B" with 7 points, and story "C" with 10 points; thus, the velocity is 21.

Velocity, Planned: This is the expected velocity for the team based on historical data. The team's actual velocity history (the average, for example) is used for planning future sprints/iterations.

Wallboard: See **Information radiator**.

Bibliography

Abrahamsson, P., Salo, O., Ronkainen, J., and Warsta, J. (2002). *Agile Software Development Methods: Review and Analysis*. Oulu, Finland: VTT Electronics.

Accenture (2015). *Next Generation Technology Delivery*. Dublin, Ireland: Accenture.

Acuna, S. T., Gomez, M., and Juristo, N. (2009). How Do Personality, Team Processes and Task Characteristics Relate to Job Satisfaction and Software Quality? *Information and Software Technology, 51* (3): 627-639. DOI: https://doi.org/10.1016/j.infsof.2008.08.006.

Ahola, J., Frühwirth, C., Helenius, M., Kutvonen, L., Myllylahti, J., Nyberg, T., Pietikäinen, A., Pietikäinen, P., Röning, J., Ruohomaa, S., Särs, C., Siiskonen, T., Vähä-Sipilä, A., and Ylimannela, V. (2014). *Handbook of Secure Agile Software Development Life Cycle*. Oulu, Finland: University of Oulu.

Al-Zewairi, M., Biltawi, M., Etaiwi, W., and Shaout, A. (2017) Agile Software Development Methodologies: Survey of Surveys. *Journal of Computer and Communications, 5* (5): 74-97. DOI: 10.4236/jcc.2017.55007.

American Management Association (2016). *The Quest for Innovation: A Global Study of Innovation Management*. New York, NY: American Management Association.

Anderson, S. P. (2011). *Seductive Interaction Design: Creating Playful, Fun, and Effective User Experiences*. Berkeley, CA: New Riders.

Andrews, A., and Bevolo, M. (2004). Understanding Digital Futures. *Design Management Journal, 15* (1): 50–57. DOI: 10.1111/j.1948-7169.2004.tb00150.

Atlas, A. (2009). Accidental Adoption: The Story of Scrum at Amazon.com. *Agile Conference, 2009*: 135-140. DOI: 10.1109/AGILE.2009.10.

Azizyan, G., Magarian, M., and Kajko-Mattson, M. (2012). The Dilemma of Tool Selection for Agile Project Management. *7th International Conference on Software Engineering Advances (ICSEA 2012)*: 605-614. ISBN: 978-1-61208-230-1.

Bass, L., Clements, P., and Kazman, R. (2012). *Software Architecture in Practice—Third Edition*. Upper Saddle River, NJ: Addison-Wesley Professional.

Beck, K., Beedle, M., van Bennekum, A., Cockburn, A., Cunningham, W., Fowler, M., Grenning, J., Highsmith, J., Hunt, A., Jeffries, R., Kern, J., Marick, B., Martin, R. C., Mellor, S., Schwaber, K., Sutherland, J. and Thomas, D. (2001). *Manifesto for Agile Software Development*. Retrieved from Manifesto for Agile Software Development: http://Agilemanifesto.org/.

Boerman, M. P., Lubsen, Z., Tamburri, D. A., and Visser, J. (2015). Measuring and Monitoring Agile Development Status. *2015 IEEE/ACM 6th International Workshop on Emerging Trends in Software Metrics, 0* (0): 54-62. DOI: 10.1109/WETSoM.2015.15.

Boehm, B. (2002). Get Ready for Agile Methods, with Care. *IEEE Computer, 32* (1): 64-69. DOI: 10.1109/2.976920.

Brown, K. A., Hyer, N. L., and Ettenson, R. (2013). The Question Every Project Team Should Answer. *MIT Sloan Management Review, 55* (1): 49-57.

CA Technologies (2015). *Agile Operations*. New York, NY: CA Technologies.

Canty, D. (2015). *Agile for Project Managers*. Boca Raton, FL: CRC Press.

CEB (2015). *Future of Government IT: Adaptive IT*. Arlington, VA: CEB, Inc.

Ceschi, M., Sillitti, A., Succi, G., and De Panfilis, S. (2005). Project Management in Plan-based and Agile Companies. *IEEE Software, 22* (3): 21-27. DOI: 10.1109/MS.2005.75.

Cho, J. (2008). Issues and Challenges of Agile Software Development with Scrum. *Journal of Management Information Systems, 9* (2): 188-195.

Chow, T., and Cao, D. (2007). A Survey Study of Critical Success Factors in Agile Software Projects. *Journal of Systems and Software, 81* (2008): 961–971. DOI: 10.1016/j.jss.2007.08.020.

Cisco (2016). *How Cisco IT Uses Agile Development with Distributed Teams and Complex Projects*. San Jose, CA: Cisco Systems, Inc.

Cisco (2011). *Agile Product Development at Cisco: Collaborative, Customer-Centered Software Development*. San Jose, CA: Cisco Systems, Inc.

Citigroup (2013). *Disruptive Innovation: Ten Things to Stop and Think About*. New York, NY: Citigroup.

Cohn, M. (2005). *Agile Estimating and Planning*. Upper Salle River, NJ: Pearson Education, Inc.

CollabNet (2014). *Agile Project Development at Intel: A Scrum Odyssey*. San Francisco, CA: CollabNet, Inc.

Conboy, K. (2009). Agility from First Principles: Reconstructing the Concept of Agility in Information Systems Development. *Information Systems Research, 20* (3): 329-354. DOI: 10.1287/isre.1090.0236.

Conforto, E. C., Salum, F., Amaral, D. C., da Silva, S. L., and Magnanini de Almeida, L. F. (2014). Can Agile Project Management Be Adopted by Industries Other than Software Development? *Project Management Journal, 45* (3): 21–34. DOI: 10.1002/pmj.21410.

DeCarlo, D. (2004). *eXtreme Project Management: Using Leadership, Principles, and Tools to Deliver Value in the Face of Volatility*. San Francisco, CA: Jossey-Bass.

Deloitte Touche Tohmatsu (2015). *Scaling Agile at Financial Institutions: Lessons from the Trenches*. New York, NY: Deloitte Touche Tohmatsu Limited.

Deloitte Touche Tohmatsu (2014). *How FrAgile is Your Agile? Six Common Pitfalls Facing Agile Project Teams*. New York, NY: Deloitte Touche Tohmatsu Limited.

Deloitte Touche Tohmatsu (2011). *Don't Fear Change, Embrace It: Advancing the Case for Agile Methods in Systems Integration*. New York, NY: Deloitte Touche Tohmatsu Limited.

Dinakar, K. (2009). *Agile Development: Overcoming a Short-Term Focus in Implementing Best Practices*. Pittsburgh, PA: Carnegie Mellon University.

Dybå, T., and Dingsøyr, T. (2008). Empirical Studies of Agile Software Development: A Systematic Review. *Information and Software Technology, 50* (9-10): 833-859.

Economist Intelligence Unit, The (2015). *Implementing the Project Portfolio: A Vital C-Suite Focus*. Newtown Square, PA: Project Management Institute, Inc.

Ernst & Young (2014). *Sustaining Digital Leadership! Agile Technology Strategies for Growth, Business Models and Customer Engagement*. London, England: Ernst & Young.

Esposito, R., Kraenzel, C. J., Pepin, C. G., and Stein, A. I. (2011). *The New Workplace: Are You Ready? How to Capture Business Value*. Somers, NY: IBM Global Services.

Fitzgerald, B., Hartnett, G., and Conboy, K. (2006). Customising Agile Methods to Software Practices at Intel Shannon. *European Journal of Information Systems, 15* (2): 200-213. DOI: 10.1057/palgrave.ejis.3000605.

Forbes (2016). *Lessons from Leaders: Marrying the Art and Science of Performance Improvement*. Jersey City, NJ: Forbes Media.

Freudenberg, S., and Sharp, H. (2010). The Top 10 Burning Research Questions from Practitioners. *IEEE Software, 27* (5): 8-9. DOI: 10.1109/MS.2010.129.

Gandomani, T. J., Zulzalil, H., Ghani, A. A. A., Sultan, A. B. M., and Nafchi, M. Z. (2013). Obstacles in Moving to Agile Software Development Methods; at a Glance. *Journal of Computer Science, 9* (5): 620-625. DOI: 10.3844/jcssp.2013.620.625.

Gartner (2014). *Taming the Digital Dragon: The 2014 CIO Agenda*. Stamford, CT: Gartner, Inc.

Gartner (2014). *Top 10 Strategic Predications for 2015 and Beyond: Digital Business is Driving Big Change*. Stamford, CT: Gartner, Inc.

Gill, M. (2013). *The Luxury Retailer's Guide to Agile Commerce*. Cambridge, MA: Forrester Research, Inc.

Glazer, H., Dalton, J., Anderson, D., Konrad, M., and Shrum, S. (2008). *CMMI or Agile: Why Not Embrace Both!* Pittsburgh, PA: Carnegie Mellon University.

Google (2015). *Design Sprint Methods: Playbook for Start Ups and Designers*. Mountain View, CA: Google.

Gorans, P., and Kruchten, P. (2014). *A Guide to Critical Success Factors in Agile Delivery*. Washington, DC: IBM Center for The Business of Government.

Graffius, S. M. (2016). *Agile in 60 Minutes: An Overview for Executives*. Los Angeles, CA: Exceptional PPM and PMO Solutions.

Graffius, S. M. (2016). *Implementing Your Value-Driven Center of Excellence PMO with a Flexible Framework to Support Agile, Traditional, and Hybrid Approaches to Development and Delivery*. Los Angeles, CA: Exceptional PPM and PMO Solutions.

Gruber, J. (2010). Apple's Constant Iterations. *Macworld, 27* (4): 100.

Harrington, H. J., Voehl, F., and Voehl, C. F. (2014). *Model for Sustainable Change*. Newtown Square, PA: Project Management Institute.

Harvard Business Review (2015). *Agile Practice: The Competitive Advantage for a Digital Age*. Boston, MA: Harvard Business Publishing.

Hewlett-Packard (2013). *Manage Projects Effectively: HP Project and Portfolio Management Center and HP Agile Manager*. Palo Alto, CA: Hewlett-Packard Development Company LP.

Hinsen, K. (2015). Technical Debt in Computational Science. *Computing in Science and Engineering, 17* (6): 103-107. DOI: 10.1109/MCSE.2015.113.

Hoda, R., Noble, J., and Marshall, S. (2013). Self-Organizing Roles on Agile Software Development Teams. *Software Engineering, 39* (3): 422-444. DOI: 10.1109/TSE.2012.30.

IBM (2014). *The IBM Agile Information Governance Process*. Somers, NY: IBM Corporation.

Institute of Practitioners in Advertising (2014). *A is for Agility*. London, United Kingdom: Institute of Practitioners in Advertising.

International Council on Systems Engineering (2004). *Systems Engineering Handbook*. Seattle, WA: International Council on Systems Engineering.

Ismail, U., Qadri, S. and Fahad, M. (2015) Requirement Elicitation for Open Source Software by Using SCRUM and Feature Driven Development. *International Journal of Natural and Engineering Sciences, 9* (1): 38-43.

Jahr, M. (2014). A Hybrid Approach to Quantitative Software Project Scheduling Within Agile Frameworks. *Project Management Journal, 45* (3): 35-45. DOI: 10.1002/pmj.21411.

Johnson, J., Boucher, K. D., Connors, K., and Robinson, J. (2001, February/March). Collaboration: Development and Management—Collaborating on Project Success. *Software Magazine, 21* (2): Supplement.

Kayes, I., Sarker, M., and Chakareski, J. (2016). Product Backlog Rating: A Case Study on Measuring Test Quality in Scrum. *Innovations in Systems and Software Engineering, 12* (4): 303-317. DOI: 10.1007/s11334-016-0271-0.

Keith, C. (2010). *Agile Game Development with Scrum*. Boston, MA: Pearson Education, Inc.

Kerzner, H. (2013). *Project Management: A Systems Approach to Planning, Scheduling, and Controlling—Eleventh Edition*. Hoboken, NJ: John Wiley & Sons, Inc.

KPMG (2015). *Moving Agility to the CIO Agenda: Towards Enterprise-Grade Agile Management*. Amstelveen, Netherlands: KPMG International.

KPMG (2011). *Agile in the IT World*. Amstelveen, Netherlands: KPMG International.

Larman, C., and Basili, V. R. (2003). Iterative and Incremental Development: A Brief History. *Computer, 36* (6): 47-56. DOI: 10.1109/MC.2003.1204375.

Larman, C., and Vodde, B. (2010). *Practices for Scaling Lean and Agile Development*. Upper Saddle River, NJ: Addison-Wesley.

Larusdottir, M. K., Cajander, A., and Gulliksen, J. (2014): Informal Feedback Rather Than Performance Measurements — User Centred Evaluation in Scrum Projects. *Behaviour and Information Technology, 33* (11): 1118-1135. DOI: 10.1080/0144929X.2013.857430.

Lukens, S., and Rogers, S. B. (2006). *The Agile CFO: Enabling the Innovation Path to Growth*. Somers, NY: IBM Corporation.

Malik, S. (2005). *Enterprise Dashboards: Design and Best Practices for IT*. Hoboken, NJ: John Wiley & Sons, Inc.

McConnell, S. (1996). *Rapid Development: Taming Wild Software Schedules*. Redmond, WA: Microsoft Press.

Medinilla, A. (2012). *Agile Management: Leadership in an Agile Environment*. New York, NY: Springer.

Moore, G. A. (2014). *Crossing the Chasm, Third Edition: Marketing and Selling Disruptive Products to Mainstream Customers*. New York, NY: HarperBusiness Publishers.

Morampudi, N. S., and Gaurav, R. (2013). Evaluating Strengths and Weaknesses of Agile Scrum Framework Using Knowledge Management. *International Journal of Computer Applications, 65* (23): 1-6. DOI: 10.5120/11221-6058.

Northrop Grumman (2013). *Health IT: Agile Project Management Office (PMO)*. Falls Church, VA: Northrop Grumman Corporation.

Oracle (2011). *Accelerate Product Innovation and Maximize Profitability: Agile Product Lifecycle Management*. Redwood Shores, CA: Oracle Corporation.

Orr, A. (2014). *Maximize the Synergies Between ITIL and DevOps*. London, United Kingdom: AXELOS.

Penton Media (2016). 15th Annual 50 Top Event Companies. *Special Events, 35* (4): 12-14.

Performance Institute, The (2013). *Program Management for Agile: Retooling Your Approach for Agile Success*. Alexandria, VA: The Performance Institute.

Perrin, R. (2008). *Real World Project Management: Beyond Conventional Wisdom, Best Practices and Project Methodologies*. Hoboken, NJ: John Wiley & Sons, Inc.

Pikkarainen, M., Haikara, J., Salo, O., Abrahamsson, P., and Still, J. (2008). The Impact of Agile Practices on Communication in Software Development. *Empirical Software Engineering, 13* (3): 303-337. DOI: 10.1007/s10664-008-9063-9.

Pohlman, W. (2017). *IEEE Computer Society/Software Engineering Institute Watts S. Humphrey Software Process Achievement (SPA) Award 2016: Nationwide (CMU/SEI-2017-TR-003)*. Retrieved from the Software Engineering Institute, Carnegie Mellon University website: http://resources.sei.cmu.edu/library/asset-view.cfm?AssetID=496270.

Poppendieck, M., and Poppendieck, T. (2003). *Lean Software Development: An Agile Toolkit*. Upper Salle River, NJ: Addison-Wesley.

Pratt, E. (2013). *Towards an Agile Authoring Methodology: Learning from the Lean Methodology*. San Jose, CA: Adobe Systems Incorporated.

Price, L. (2014). People Leadership: From Control to Vision. *CIO New Zealand, 17* (4): 6.

PricewaterhouseCoopers (2014). *Adopting and Agile Methodology: Requirements-gathering and Delivery*. London, United Kingdom: PricewaterhouseCoopers LLP.

PricewaterhouseCoopers (2013). *Scale Agile Throughout the Enterprise: A PwC Point of View*. London, United Kingdom: PricewaterhouseCoopers LLP.

Project Management Institute (2014). *Change Readiness: Focusing Change Management Where It Counts*. Newtown Square, PA: Project Management Institute.

Project Management Institute (2013). *A Guide to the Project Management Body of Knowledge—Fifth Edition*. Newtown Square, PA: Project Management Institute.

Project Management Solutions (2014). *The State of the Project Management Office (PMO): 2014*. Glen Mills, PA: Project Management Solutions, Inc.

Protiviti (2015). *Agile Technology Controls for Startups—a Contradiction in Terms or a Real Opportunity?* New York, NY: Protiviti, Inc.

Qumer, A., and Henderson-Sellers, B. (2008) An Evaluation of the Degree of Agility in Six Agile Methods and Its Applicability for Method Engineering. *Information and Software Technology, 50* (4): 280-295. https://doi.org/10.1016/j.infsof.2007.02.002.

Ramadan, N., and Megahed, S. (2016). Requirements Engineering in Scrum Framework. *International Journal of Computer Applications, 149* (8): 24-29. DOI: 10.5120/ijca2016911530.

Robinson, P. K., and Hsieh, L. (2016). Reshoring: A Strategic Renewal of Luxury Clothing Supply Chains. *Operations Management Research, 9* (3): 89-101. DOI: 10.1007/s12063-016-0116-x.

Rising, L., and Janoff, N. S. (2000). The Scrum Software Development Process for Small Teams. *IEEE Software, 17* (4): 26-32. DOI: 10.1109/52.854065.

Sampaio, A., Vasconcelos, A., and Sampaio, P. R. F. (2004). Assessing Agile Methods: An Empirical Study. *Journal of the Brazilian Computer Society, 10* (2): 21-48. DOI: 10.1007/BF03192357.

Schneiderman, M. (2013). *Digital Dashboards: Best Practices and Lessons Learned.* Princeton, NJ: Information Age Associates.

Schwaber, K. (2006). *The Enterprise and SCRUM: Best Practices.* Redmond, WA: Microsoft Press.

Scrum Alliance (2013). *The State of Scrum: Benchmarks and Guidelines.* Newtown Square, PA: ProjectManagement.com.

Sharif, B., Khan, S. A., and Bhatti, M. W. (2012). Measuring the Impact of Changing Requirements on Software Project Cost: An Empirical Investigation. *International Journal of Computer Science Issues, 9* (3): 170-174.

Solinski, A., and Petersen, K. (2016). Prioritizing Agile Benefits and Limitations in Relation to Practice Usage. *Software Quality Journal, 24* (2): 447-482. DOI: 10.1007/s11219-014-9253-3.

Strålin, T., Gnanasambandam, C., Andén, P., Comella-Dorda, S., and Burkacky, O. (2016). *Software Development Handbook: Transforming for the Digital Age.* Stockholm, Sweden: McKinsey & Company, Inc.

Sutherland, J., Harrison, N., and Riddle, J. (2014). Teams that Finish Early Accelerate Faster: A Pattern Language for High Performing Scrum Teams. Presented at 2014 47th Hawaii International Conference on System Science. *System Sciences*: 4722-4728. DOI: 10.1109/HICSS.2014.580.

Sutherland, R. (2011). *Rory Sutherland: The Wiki Man.* London, United Kingdom: It's Nice That, and Ogilvy Group Ltd.

Takeuchi, H., and Nonaka, I. (1986). The New New Product Development Game. *Harvard Business Review, 64* (1): 137-146.

Technology Business Management Council (2014). *DirecTV: Fail Fast, Innovate Often*. Bellevue, WA: Technology Business Management Council.

Thompson, K. (2011). *Agile for Executives*. Foster City, CA: cPrime, Inc.

Vatier, C. (2015). *Incubator or Respirator? Why You Need to Change the Way You Innovate. Now*. Dublin, Ireland: Accenture.

VersionOne (2016). *10th Annual State of Agile Survey*. Atlanta, GA: VersionOne, Inc.

Vijay, D., and Ganapathy, G. (2014). Guidelines to Minimize the Cost of Software Quality in Agile Scrum Process. *International Journal of Software Engineering and Applications, 5* (3). DOI: 10.5121/ijsea.2014.5305.

Wake, B. (2003). *INVEST in Good Stories, and SMART Tasks*. Retrieved at http://xp123.com/articles/invest-in-good-stories-and-smart-tasks/.

Wale-Kolade, A., and Nielsen, P. A. (2016). Apathy towards the Integration of Usability Work: A Case of System Justification. *Interacting with Computers, 28* (4): 437-450. DOI: 10.1093/iwc/iwv016.

Weill, P., and Ross, J. W. (2004). *IT Governance on One Page*. Cambridge, MA: Massachusetts Institute of Technology.

West, D. (2011). *Water-Scrum-Fall is the Reality of Agile for Most Organizations Today*. Cambridge, MA: Forrester Research, Inc.

Whitaker, S. (2014). *The Benefits of Tailoring: Making a Project Management Methodology Fit*. Newtown Square, PA: Project Management Institute, Inc.

Appendices

This additional content is succinct, usually about a page per topic

The appendix is followed by three sections:

Agile Manifesto

In February of 2001, a group of software engineers met and brainstormed ways to improve software development. They developed *The Agile Manifesto*.

The Agile Manifesto

We are uncovering better ways of developing software by doing it and helping others do it. Through this work we have come to value:

- Individuals and interactions over processes and tools

- Working software over comprehensive documentation

- Customer collaboration over contract negotiation

- Responding to change over following a plan

That is, while there is value in the items on the right, we value the items on the left more.

"Items on the left" refers to the content to the left of the word "over." For example, "following a plan" is valuable but "responding to change" is more valuable.

The original manifesto authors included Kent Beck, Alistair Cockburn, James Grenning, Ron Jeffries, Robert C. Martin, Jeff Sutherland, Mike Beedle, Ward Cunningham, Jim Highsmith, Jon Kern, Steve Mellor, Dave Thomas, Arie van Bennekum, Martin Fowler, Andrew Hunt, Brian Marick, and Ken Schwaber.

This section on Agile Manifesto uses material from the Wikipedia article https://en.wikipedia.org/wiki/Agile_software_development, which is released under the Creative Commons Attribution-Share-Alike License (http://creativecommons.org/licenses).

There's supplemental information in the following section of the appendix:

Agile Principles

Agile Models

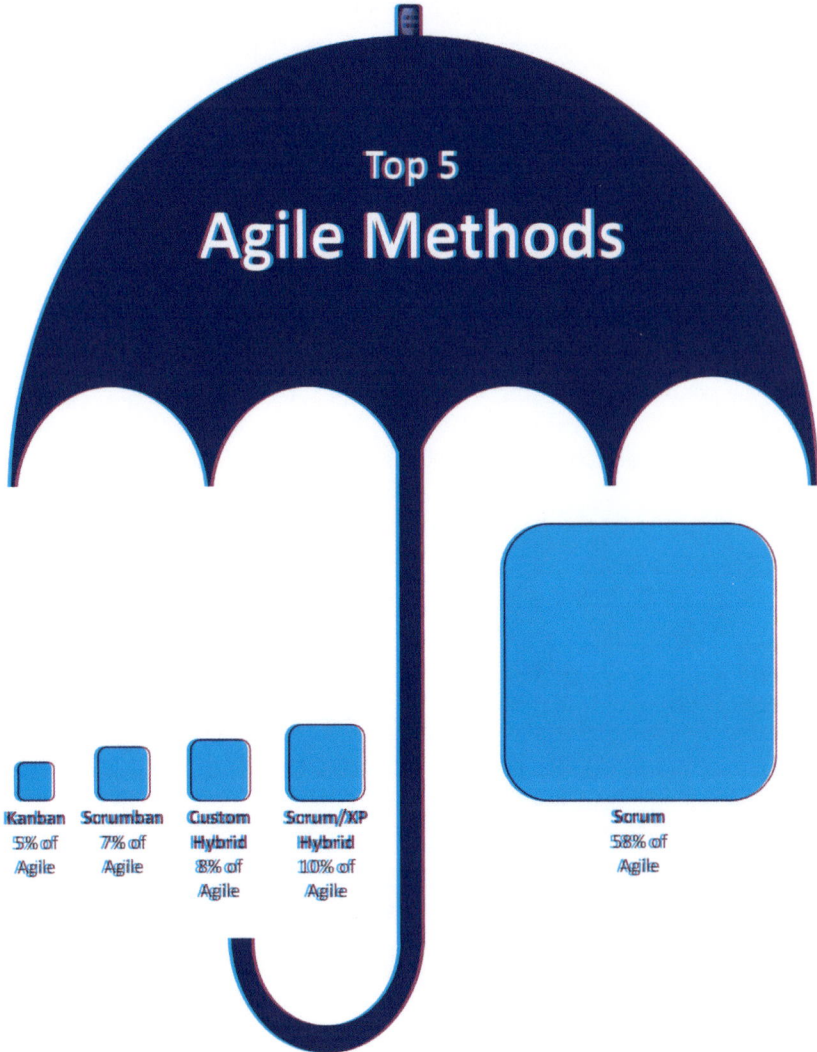

Top 5
Agile Methods

Kanban
5% of
Agile

Scrumban
7% of
Agile

Custom
Hybrid
8% of
Agile

Scrum/XP
Hybrid
10% of
Agile

Scrum
58% of
Agile

Source of data: VersionOne (2016). *10th Annual State of Agile Survey*. Atlanta, GA: VersionOne, Inc.

Additional agile models—not in the top five above—include: Adaptive software development, Agile Unified Process, Crystal Clear methods, disciplined agile delivery, dynamic systems development method, Extreme programming, feature-driven development, lean software development, and rapid application development.

Agile Principles

In 2001, a group of software engineers met and brainstormed ways to improve software development. They developed *The Agile Manifesto*. As a follow up to the manifesto, they established the following philosophies.

Twelve Principles of Agile Software

1. Our highest priority is to satisfy the customer through early and continuous delivery of valuable software.

2. Welcome changing requirements, even late in development. Agile processes harness change for the customer's competitive advantage.

3. Deliver working software frequently, from a couple of weeks to a couple of months, with a preference to the shorter timescale.

4. Business people and developers must work together daily throughout the project.

5. Build projects around motivated individuals. Give them the environment and support they need, and trust them to get the job done.

6. The most efficient and effective method of conveying information to and within a Development Team is face-to-face conversation.

7. Working software is the primary measure of progress.

8. Agile processes promote sustainable development. The sponsors, developers, and users should be able to maintain a constant pace indefinitely.

9. Continuous attention to technical excellence and good design enhances agility.

10. Simplicity – the art of maximizing the amount of work not done – is essential.

11. The best requirements and designs emerge from self-organizing teams.

12. At regular intervals, the team reflects on how to become more effective, then tunes and adjusts its behavior accordingly.

This section on Agile Principles uses material from the Wikipedia article https://en.wikipedia.org/wiki/Agile_software_development, which is released under the Creative Commons Attribution-Share-Alike License (http://creativecommons.org/licenses).

Agile Scrum Overview

Agile is a group of methodologies that promote adaptive planning, evolutionary development, and incremental delivery. Scrum is the leading agile development and project delivery framework. It's a powerful set of principles and practices that help teams deliver products in short cycles, as well as enable speedy feedback, rapid adaptation to change, fast delivery time, and continuous improvement.

Key principles of the Scrum framework include:

- Emphasis on transparency, inspection, and adaptation

- Self-organized teams for better buy-in and shared ownership

- Collaboration for a shared value-creation process

- Value-based prioritization to deliver maximum business value

- Time-boxing to manage planning and execution

- Iterative development as a means to better manage change and satisfy customer needs

Scrum enables organizations to adjust to rapidly-changing requirements and produce a product that meets evolving business objectives.

That's it for this quick look. If you'd like to learn more about Agile Scrum, you'll find additional content in the **Appendix**.

There's supplemental information in the following sections of the appendix:

Agile Manifesto   Agile Principles   Declaration of Interdependence

Project Assessment Tool   When to Use Agile Scrum

Artifacts Overview

Scrum's artifacts are designed to provide transparency of key information and opportunities for inspection and adaptation. Here's an overview of the artifacts.

Product Vision

- One or two sentence summary
- Declared by Product Owner
- Accepted by the Development Team

Sprint Burndown Chart

- Displays progress for a sprint
- Shows how much remains

Product Release Backlog

- List of features, requirements, bugs
- Contains estimates (in story points)
- Prioritized by the Product Owner

Product Release Burndown Chart

- Displays progress for a product release
- Shows how much work remains

Sprint Goal

- One sentence summary for the sprint
- Declared by the Product Owner
- Accepted by the Development Team

Impediment Backlog

- Tool to track problems through resolution
- Owned by the Scrum master

Sprint Backlog

- Work committed to be done in a sprint
- Comprised of well understood stories and tasks

Potentially Shippable Increment

- Item from Sprint Backlog that's "Done"
- Implemented, tested, free of defects, etc.
- Capable of being shipped

The most effective artifacts provide "just enough" structure and control. Questions to help assess value include:

- Are they easy to use, read, and understand?

- Are they kept updated?

- Are they readily available/easily accessed?

- Are they helpful?

The preceding was a high-level presentation on Scrum artifacts. They're covered further in other sections of this guide.

Communications Management Comparison

Comparisons between communications management in Agile Scrum and traditional models (such as waterfall) are highlighted below.

Communications Management Overview		
Agile Scrum Model		**Traditional Model**
• Follow Scrum framework, which includes Scrum team, events (formerly known as ceremonies), and artifacts • Generally, informal (face-to-face) communications are favored	≈ APPROXIMATELY EQUIVALENT TO	• Identify stakeholders • Plan communications • Generally, formal (written) communications are favored
• Product rollout and release plan • Product release backlog • Sprint planning meeting, Part 1 • Sprint planning meeting, Part 2 • Sprint backlog • Task board • Sprint burndown chart • Impediment backlog • Information radiator • Daily Scrum (daily stand-up meetings) • Sprint review meeting • Sprint retrospective meeting	≈ APPROXIMATELY EQUIVALENT TO	• Distribute information • Manage expectations • Report performance

There's supplemental information in the following sections of the appendix:

Cost Management Comparison

Human Resources/Team Management Comparison

Integration Management Comparison Quality Management Comparison

Risk Management Comparison   Scope Management Comparison

Time Management Comparison

Cost Management Comparison

Comparisons between cost management in Agile Scrum and traditional models (such as waterfall) are presented below.

Cost Management Overview		
Agile Scrum Model		**Traditional Model**
• Estimate costs for proposed solution • Sponsor approves funding • Iterative value-driven development	APPROXIMATELY EQUIVALENT TO	• Estimate costs • Determine budget • Control costs

There's supplemental information in the following sections of the appendix:

Communications Management Comparison

Human Resources/Team Management Comparison

Integration Management Comparison   Quality Management Comparison

Risk Management Comparison   Scope Management Comparison

Time Management Comparison

Declaration of Interdependence

Authored by a consortium of project managers, *The Declaration of Interdependence* serves as a supplement to *The Agile Manifesto*. It lists management values needed to reinforce an agile development mentality. It's also known as *The Declaration of Interdependence for Modern Management*.

The Declaration of Interdependence

We increase return on investment by making continuous flow of value our focus.

We deliver reliable results by engaging customers in frequent interactions and shared ownership.

We expect uncertainty and manage for it through iterations, anticipation and adaptation.

We unleash creativity and innovation by recognizing that individuals are the ultimate source of value and creating an environment where they can make a difference.

We boost performance through group accountability for results and shared responsibility for team effectiveness.

We improve effectiveness and reliability through situationally specific strategies, processes and practices.

The authors included David Anderson, Sanjiv Augustine, Christopher Avery, Alistair Cockburn, Mike Cohn, Doug DeCarlo, Donna Fitzgerald, Jim Highsmith, Ole Jepsen, Lowell Lindstrom, Todd Little, Kent McDonald, Pollyanna Pixton, Preston Smith, and Robert Wysocki.

This section on Declaration of Interdependence uses material from the Wikipedia article https://en.wikipedia.org/wiki/PM_Declaration_of_Interdependence, which is released under the Creative Commons Attribution-Share-Alike License (http://creativecommons.org/licenses).

There's supplemental information in the following sections of the appendix:

Agile Manifesto   Agile Principles

Definition of Done

A clear definition of done (DoD) is attached to all work being done in the sprint. It provides precision about when work is complete. Work is marked done only after it meets the DoD benchmark.

Here's a basic example:

- Code complete

- Unit tests written and executed

- Integration tested

- Performance tested

- Documented

- Acceptance criteria met

- Accepted by Product Owner

Development Team

In Scrum, there are three core roles: Product Owner, Scrum Master, and Development Team. Together they are the Scrum team. The Development Team member role is featured here.

The Development Team is a cross-functional group (which often includes domain experts, business analysts, developers, testers) responsible for delivering potentially shippable increments of product at the end of every sprint. The Development Team is comprised of between three and nine (a prior guideline noted between five and nine) people.

Responsibilities

- Self-organizing

- Negotiating commitments with the Product Owner

- Collaborating with anyone necessary to get the job done

- Delivering the product

- Alerting the Scrum Master of any impediments

- Presenting the product at the sprint review (demo)

- Inspecting and adapting the process

Considerations

- Stakeholder interests

- Commitments made to the Product Owner

Decisions

- Implementation details

There's supplemental information in the following sections of the appendix:

Self-Organization   Scrum Master   Product Owner   Stakeholders

Testing

Events Overview

Here's an overview of the four key sprint events (meetings). Events were previously called ceremonies.

Sprint Planning

- Identify which product release backlog items will be delivered (Part 1)
- Determine how the work will be achieved (Part 2)
- Time-box: Two hours for each week in the sprint

Daily Scrum

- Daily stand-up meeting
- Covers progress since last meeting, work planned for today, and any impediments
- Time-box: 15 minutes

Sprint Review

- Live demo of working "done" functionality
- Time-box: One hour for each week in the sprint

Sprint Retrospective

- Identify what went well, what didn't go well, and how the team can improve in the next sprint
- Time-box: 45 minutes for each week in the sprint

The preceding was a quick look at the events. They're covered further in other sections of this guide.

Fibonacci Sequence and Story Points

The Fibonacci sequence—attributed to Italian mathematician Leonardo Pisano—is a series of numbers where a value is found by adding up the two numbers before it. Starting with 0 and 1, the sequence goes 0, 1, 1, 2, 3, 5, 8, 13, 21, 34, etc.

0, 1, 1, 2, 3, 5, 8, 13, 21, 34, ...

0 + 1 = 1 2 3 5 8 13 21 34 ...

0 **1 + 1 = 2** 3 5 8 13 21 34 ...

0 1 **1 + 2 = 3** 5 8 13 21 34 ...

0 1 1 **2 + 3 = 5** 8 13 21 34 ...

0 1 1 2 **3 + 5 = 8** 13 21 34 ...

0 1 1 2 3 **5 + 8 = 13** 21 34 ...

0 1 1 2 3 5 **8 + 13 = 21** 34 ...

0 1 1 2 3 5 8 **13 + 21 = 34** ...

If you're interested in the mathematical expression, the numbers are generated by setting $F_0=0$, $F_1=1$, and then using the recursive formula $F_n = F_{n-1} + F_{n-2}$.

In Agile Scrum, story points are often used to rate the complexity of a user story (work). Most teams use the Fibonacci sequence or a variation of it as an estimating technique. How-to details are provided in **Section 4B: Estimation via Story Points**.

Human Resources/Team Management Comparison

Comparisons between human resources/team management in Agile Scrum and traditional models (such as waterfall) are highlighted below.

Human Resources/Scrum Team Management Overview

Agile Scrum Model		Traditional Model
• Follow Scrum framework inclusive of Scrum team totaling between five and eleven (formerly between seven and eleven) people: one Product Owner, one Scrum Master, and between three and nine (formerly between five and nine) Development Team members	APPROXIMATELY EQUIVALENT TO	• Develop plan • Acquire team – can be any composition and size • Develop team
• Scrum teams are self-directed – free to determine how deliverables are fulfilled	APPROXIMATELY EQUIVALENT TO	• Manage team – individuals are generally prescribed how work is done

There's supplemental information in the following sections of the appendix:

Communications Management Comparison

Cost Management Comparison   Integration Management Comparison

Quality Management Comparison   Risk Management Comparison

Scope Management Comparison   Time Management Comparison

Information Radiator

An information radiator—sometimes referred to as a wallboard or an electronic wallboard—is a large display of vital information that is continuously updated and stationed where it can be seen by the Scrum team and stakeholders. It's a highly valuable tool for Scrum teams to share information and maintain focus on common goals.

Information radiators can take different forms, ranging from information hand-drawn on flip chart paper or whiteboards to the use of software tools. If you're new to Scrum, you may find it best to begin with flip chart paper or a whiteboard. Later, can consider an electronic wallboard by using commercial off-the-shelf software or by building a custom solution (an example follows).

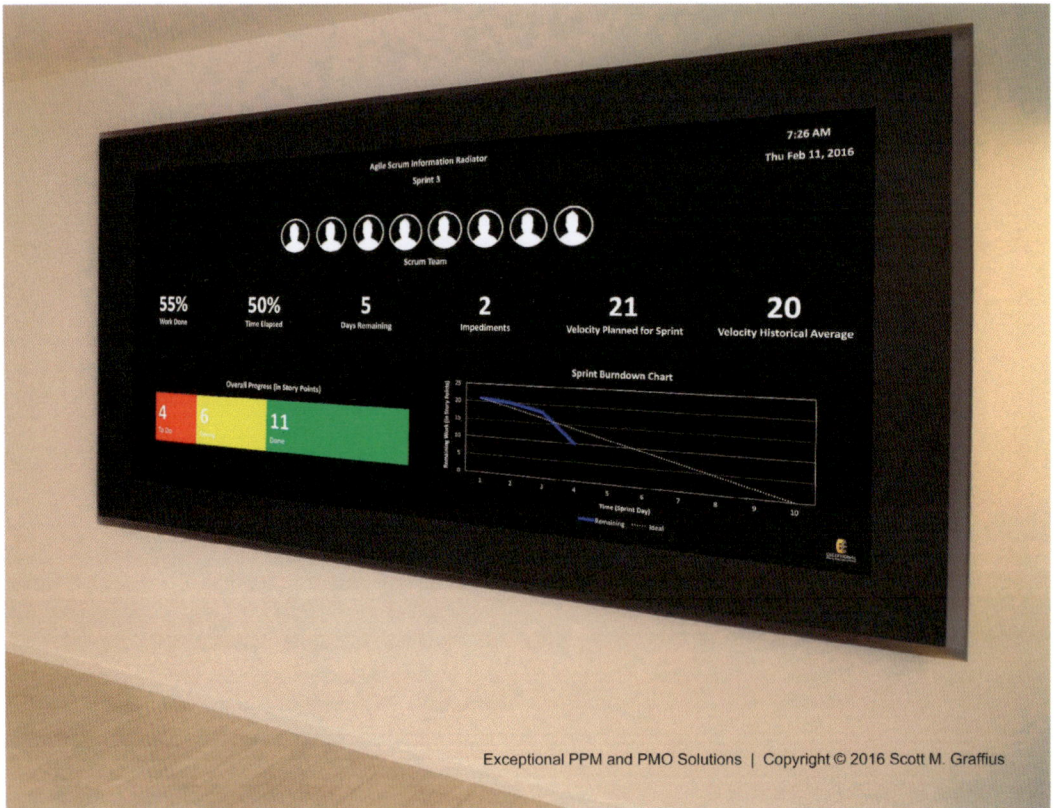

Exceptional PPM and PMO Solutions | Copyright © 2016 Scott M. Graffius

The image above displays a custom information radiator on a very large monitor. For privacy and confidentiality, it is intentionally shown without branding for the client organization and without profile pictures for the Scrum team.

> There's supplemental information in the following section of the appendix:
>
> Metrics

Inspect and Adapt

Cycles in Scrum are relatively short by design, and each one provides the team with multiple opportunities to assess status, receive feedback, and learn. There are four formal events (meetings) for inspection and adaptation:

- Sprint Planning—detailed in **Section 5: Sprint Planning Part 1** and **Section 6: Sprint Planning Part 2 (Sprint Backlog)**

- Daily Scrum—covered in **7D: Daily Scrum Meeting**

- Sprint Review—described in **Section 8: Sprint Review**

- Sprint Retrospective—covered in **Section 9: Sprint Retrospective**

Information learned in the current cycle informs planning and changes for the next one. That's inspect and adapt in Agile Scrum.

Integration Management Comparison

Comparisons between integration management in Agile Scrum and traditional models (such as waterfall) are presented below.

Integration Management Overview		
Agile Scrum Model		**Traditional Model**
• Product vision • Sprint goal	≈ APPROXIMATELY EQUIVALENT TO	• Project charter development
• Product release backlog • Sprint review meeting, Part 1 • Sprint review meeting, Part 2 • Sprint backlog	≈ APPROXIMATELY EQUIVALENT TO	• Project plan development
• Scrum core roles: Scrum Master, Product Owner, and Development Team • Scrum events (formerly called ceremonies) • Scrum artifacts	≈ APPROXIMATELY EQUIVALENT TO	• Project management, monitoring, and control
• Continuous feedback • Prioritized product release backlog • Prioritized sprint backlog	≈ APPROXIMATELY EQUIVALENT TO	• Integrated change control

There's supplemental information in the following sections of the appendix:

Communications Management Comparison

Cost Management Comparison

Human Resources/Team Management Comparison

Quality Management Comparison   Risk Management Comparison

Scope Management Comparison   Time Management Comparison

Management Support

Scrum is a very different way of managing projects for many organizations. Strong executive commitment is a success factor for implementing Scrum, and management can best demonstrate their support of the transformation through their actions.

The image below portrays the convergence of support from executives. The leadership positions are provided as an example. The number and titles of executives vary by organization.

There's supplemental information in the following sections of the appendix:

Agile Scrum Overview   Agile Manifesto   Agile Principles

Declaration of Interdependence   When to Use Agile Scrum

Metrics

The use of metrics varies. Some organizations with Agile Scrum implementations use very few measures, while others use many. General guidelines follow.

- Individuals and interactions should be more important than metrics

- Applying the agile principle of "inspect and adapt," metrics should be used and adjusted as appropriate to your unique circumstances

- Use multiple metrics

  - If you don't collect any metrics, you're flying blind

  - If you collect and focus on too many, they may be obstructing your field of view

Commonly used metrics are velocity, sprint burndown, and product release burndown.

- Velocity: The separate section of the appendix titled **Velocity** provides details on the subject

- Sprint burndown: **Section 7B: Sprint Burndown Chart** provides details

- Product release burndown: The separate section of the appendix titled **Product Release Burndown Chart** provides details on the topic

Select additional metrics—including some less frequently used measures—follow.

- Scrum team member satisfaction

  - Team members' experience with the organization, work environment, processes, and fellow team members

    - One simple technique involves the selection of a smiley face (🙂), a neutral face (😐), or a frown face (🙁)

      - If a team member picks a frown face or a neutral face, they're asked to provide comments

    - Alternatively, satisfaction can be rated on a numerical scale—such as 1-5, where 1 indicates disappointed, and 5 represents delighted

      - If a team member picks 1, 2 or 3, they're asked to provide comments

  - Any concerns should be addressed quickly

  - Satisfaction should be tracked over time

- Customer or stakeholder satisfaction

    o The techniques described under Scrum team member satisfaction can be used

- Delivered done defect density

    o Number of defects identified after work was done, divided by the size of work done in story points

    o This metric relates to the effectiveness of review and testing activities

    o Metric should be tracked over time

- Actual stories completed vs. committed stories

    o Number of stories completed divided by the number of stories committed to being done

        ▪ For the number of stories completed, use the stories identified as done in **Section 8: Sprint Review**

        ▪ For the number of stories committed to being done, use the number of stories committed to in **Section 5: Sprint Planning Part 1** and **Section 6: Sprint Planning Part 2 (Sprint Backlog)**

    o Metric should be tracked over time

- Sprint slippage risk indicator for story points

    o Historical average story points completed per week velocity, divided by story points committed to being completed per week

    o Calculated values

        ▪ Value of 1.00 reflects that the work planned matches historical productivity

        ▪ Less than 1.00 (or less than 0.80 to factor for variability) suggests that too much work is planned

        ▪ Greater than 1.00 (or more than 1.20 to factor fluctuations) indicates that not enough work is planned

    o Metric should be tracked over time

Additional information is available at scottgraffius.com/agile-scrum-bonus-material.html.

Product Owner

In Scrum, there are three core roles: Product Owner, Scrum Master, and Development Team. Together they are the Scrum team. The Product Owner role is featured here.

The Product Owner represents the interests of stakeholders, maintains the product release backlog, and ensures the value of the work done by the Development Team.

Responsibilities

- Representing the voice of customers

- Communicating with stakeholders to ensure their interests are represented

- Managing stakeholder expectations

- Establishing and achieving the product vision

- Defining releases, their goals, and sprint goals

- Managing the return on investment

- Creating and maintaining the product backlog

- Authoring and prioritizing user stories based on business value

- Outlining acceptance criteria

- Attending the sprint reviews and planning sessions

- Re-prioritizing the product backlog (continuously)

Considerations

- Stakeholder interests

Decisions

- Accepting/rejecting work

- Deciding whether to ship

- Able to cancel (abnormally terminate) a sprint if appropriate

There's supplemental information in the following sections of the appendix:

Scrum Master   Development Team   Stakeholders

Product Release Burndown Chart

Sprints and product releases each have their own burndown chart. The sprint burndown chart is covered in **7B: Sprint Burndown Chart**, and information on the product release burndown chart follows.

A product release usually spans multiple sprints, and the product release burndown chart involves:

- Horizontal x-axis displays sprints

    o   Such as sprint 1 through sprint 7

- Vertical y-axis displays remaining work

    o   In story points (this is most common), or

    o   In hours

- Ideal velocity is included as a guideline

> There's supplemental information in the following section of the appendix:
>
> Velocity

- Actual progress-to-date

These charts can be hand-drawn on flip chart paper or whiteboards, or software tools—such as those listed in **Select Resources**—can be used. If you're new to Agile Scrum, you may find it best to begin with the use of flip chart paper or a whiteboard.

The Product Owner updates the product release burndown chart at the end of each sprint.

An example of a product release burndown chart is shown next.

Product Release Burndown Chart

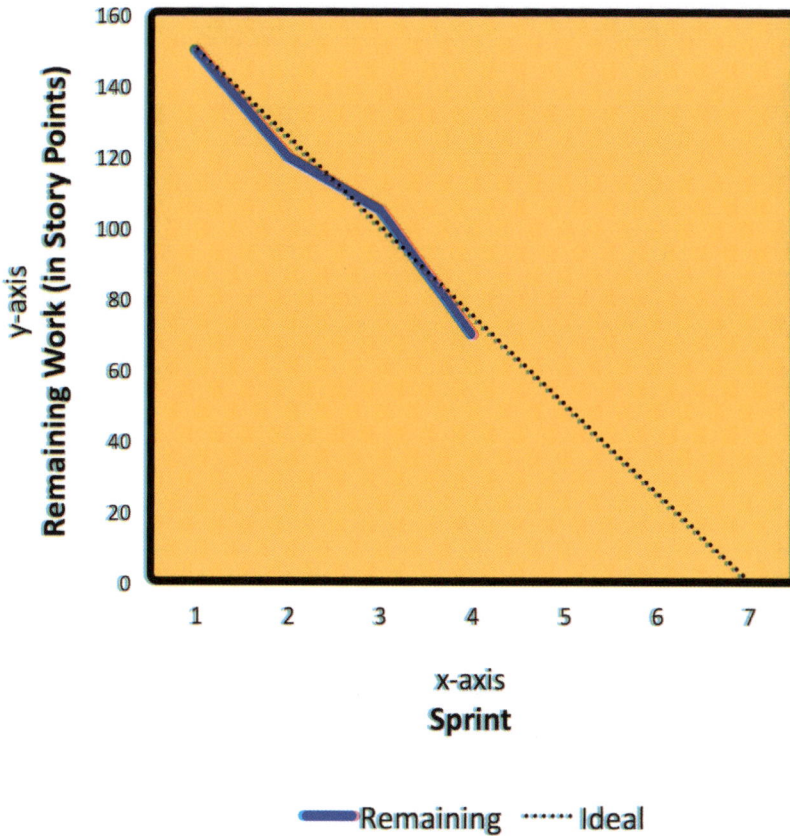

There's supplemental information in the following section of the appendix:

Metrics

Project Assessment Tool

If you have a particular project in mind, you can use this simple project assessment tool as a guide to help determine which framework—Agile Scrum, traditional, or hybrid—is most appropriate.

		Select Either	
	Project Assessment Tool © Scott M. Graffius		
	Item Description	**More Likely**	**Less Likely**
1	Project delivery and ROI can be incremental.	☐	☐
2	Core team can be at the same location.	☐	☐
3	Core team can be dedicated (assigned 100%) to the project.	☐	☐
4	Governance model allows for adaptive planning, iterative development and incremental releases.	☐	☐
5	Stakeholders are highly committed to the project.	☐	☐
6	An automated test environment is available.	☐	☐
7	Scope and/or requirements are unclear and/or unstable.	☐	☐
8	The project involves development on a single system or a single system which interfaces with others.	☐	☐
9	Continuous integration is highly possible.	☐	☐
10	Project has low dependencies on vendors or other projects.	☐	☐

Instructions

1. For each of the 10 items above, select either More Likely or Less Likely for a particular project that you have in mind.
2. Calculate the score: How many items were marked as More Likely? That's the score.
3. What it means:
   - If the score is between 8-10, the project is likely well suited for Agile Scrum
   - If the score is between 4-7, an Agile Scrum, traditional or hybrid model may apply
   - If the score is between 0-3, the project is likely well suited for a traditional model (such as waterfall)

🔗 A PDF of this tool is available at scottgraffius.com/agile-scrum-bonus-material.html.

Quality Management Comparison

Comparisons between quality management in Agile Scrum and traditional models (such as waterfall) are highlighted below.

Quality Management Overview		
Agile Scrum Model		**Traditional Model**
• Definition of done • Acceptance criteria	APPROXIMATELY EQUIVALENT TO	• Quality planning
• Test execution • Sprint review meeting • Sprint retrospective meeting	APPROXIMATELY EQUIVALENT TO	• Quality assurance
• Test results • Acceptance	APPROXIMATELY EQUIVALENT TO	• Quality control

There's supplemental information in the following sections of the appendix:

Communications Management Comparison   Cost Management Comparison

Human Resources/Team Management Comparison

Integration Management Comparison   Risk Management Comparison

Scope Management Comparison   Time Management Comparison

Risk Management Comparison

Comparisons between risk management in Agile Scrum and traditional models (such as waterfall) are presented below.

Risk Management Overview

Agile Scrum Model

- Sprint planning meeting, Part 1 and Part 2
- Impediment backlog
- Daily Scrum (daily stand-up meetings)
- Sprint retrospective meeting

APPROXIMATELY EQUIVALENT TO

Traditional Model

- Risk identification
- Risk analysis
- Risk response planning

- Impediment backlog
- Daily Scrum (daily stand-up meetings)
- Information radiator

APPROXIMATELY EQUIVALENT TO

- Risk monitoring
- Risk control

There's supplemental information in the following sections of the appendix:

Communications Management Comparison   Cost Management Comparison

Human Resources/Team Management Comparison

Integration Management Comparison   Quality Management Comparison

Scope Management Comparison   Time Management Comparison

Scope Management Comparison

Comparisons between scope management in Agile Scrum and traditional models (such as waterfall) are highlighted below.

Scope Management Overview		
Agile Scrum Model		**Traditional Model**
• Product release backlog • Sprint planning meeting, Part 1 • Sprint planning meeting, Part 2	≈ APPROXIMATELY EQUIVALENT TO	• Requirements collection • Scope definition
• Product roadmap and release plan • Product release backlog • Sprint backlog	≈ APPROXIMATELY EQUIVALENT TO	• Scope delineation – work breakdown structure
• Definition of Done • Acceptance	≈ APPROXIMATELY EQUIVALENT TO	• Scope verification
• Continuous feedback • Prioritized product release backlog • Prioritized sprint backlog	≈ APPROXIMATELY EQUIVALENT TO	• Scope change control

There's supplemental information in the following sections of the appendix:

Communications Management Comparison   Cost Management Comparison

Human Resources/Team Management Comparison

Integration Management Comparison   Quality Management Comparison

Risk Management Comparison   Time Management Comparison

Scrum Master

In Scrum, there are three core roles: Product Owner, Scrum Master, and Development Team. Together they are the Scrum team. The Scrum Master role is featured here.

The Scrum Master ensures that the team adheres to Scrum practices, values, and rules. The Scrum Master also facilitates the removal of impediments and helps protect the team from disruption.

Responsibilities

- Serving as single point of contact for the project

- Facilitating and enforcing the Scrum process

- Coaching the team on Scrum values and practices

- Facilitating the daily Scrum (stand-up meeting)

- Ensuring the team is fully functional and productive

- Helping the team remove impediments

- Capturing data

- Keeping Scrum artifacts (charts, etc.) current and visible

- Enforcing time-boxes

- Promoting improved engineering practices

- Shielding the team from external interferences and distractions

- Conducting the sprint retrospective at the end of a sprint

Considerations

- Stakeholder interests

- Commitments made to the Product Owner

Decisions

- Deciding on environment specifics that are conducive to the team and the Scrum process

- Implementation details

There's supplemental information in the following sections of the appendix:

Product Owner | Development Team | Stakeholders

Self-Organization

Self-organization plays an important role in Agile Scrum. Self-organizing teams determine how to best accomplish their work, rather than being directed by others outside the team. They are successful when:

- Team members are committed to clear, short-term goals

- Team members have a view into each other's contribution

- Team members can easily see the progress of the group

- Team members feel free to provide feedback

Characteristics of effective self-organized teams include:

- Development Team size of between three and nine (formerly between five and nine) people

- Autonomous working practice

- High level of collaboration

- Continuously improve and learn

- Cross-training of team members

Self-organized teams provide the following benefits:

- Shared ownership

- Innovative environment conducive to satisfaction and growth

- Enhanced performance

Stakeholders

Stakeholders are those not in the Scrum team but have an interest or stake in the project. Stakeholders may be direct managers of the core Scrum team, executives, users or others.

Responsibilities

- Communicating their needs

- Working with the Product Owner to develop and refine the product release backlog

- Providing feedback to the Scrum team during the sprint review meetings

- Helping to remove impediments

Considerations

- Avoid distracting the Scrum team during a sprint

- Support the Scrum framework

Decisions

- A stakeholder serving as the sponsor typically makes funding decisions

There's supplemental information in the following sections of the appendix:

Scrum Master   Product Owner   Development Team

Technical Debt

Technical debt is a term used to describe the eventual consequences of a technical design or development choice made for a short-term benefit. An example is writing suboptimal code to meet a deadline, knowing that the code will have to be rewritten later to make the software maintainable.

Technical debt may have one or more causes. Seven common sources include:

1. Time pressures

2. Overly complex technical design

3. Poor alignment to standards

4. Lack of skill

5. Suboptimal code

6. Delayed refactoring

7. Insufficient testing

Over time, those factors result in the accumulation of technical inefficiencies that need to be serviced in the future. Unchecked technical debt may make the software more expensive to change than to re-implement.

Technical debt can be avoided or minimized by not taking shortcuts, using simple designs, and refactoring continuously. When there's technical debt, the team should make the items visible by registering entries in the product release backlog, where the matters will be evaluated and prioritized for resolution.

Testing

An Agile Scrum test strategy provides structure on quality testing, and it usually has a mission statement related to business objectives. For example:

> "We constantly deliver working software that meets customer requirements by providing fast feedback and defect prevention."

The approach includes testing performed early and often, and close cooperation with developers and customers/users. The types of tests typically involved are summarized below.

	Who	What	When	Why
Unit Testing	Developers and technical architects	All new code and refactoring of legacy code	As soon as new code is written	To ensure code is developed correctly
Application Program Interface (API) and Service Testing	Developers and technical architects	New web services, components, controllers, etc.	As soon as a new API is developed and ready	To ensure communication between components are working correctly
Acceptance Testing	Developers and quality assurance	Verifying acceptance tests on user stories and verification of features	When feature is ready and unit tested	To ensure customer expectations are met
System Testing, Regression Testing (detecting any side effects from changes) and User Acceptance Testing (UAT)	Developers, quality assurance, business analysts, and product owners	Scenario testing, user flows, performance testing and security testing	When Acceptance Testing is completed	To ensure the whole system works when integrated

Tests should be automated as much as possible. Benefits include time saved (compared with executing tests manually), and automation can lead to more exhaustive testing.

Time Management Comparison

Comparisons between time management in Agile Scrum and traditional models (such as waterfall) are presented below.

Time Management Overview		
Agile Scrum Model		**Traditional Model**
• Product rollout and release plan • Product release backlog • Sprint planning meeting, Part 1 • Sprint planning meeting, Part 2 • Sprint backlog	APPROXIMATELY EQUIVALENT TO	• Activities definition • Activities sequencing • Resource and duration estimating
• Product rollout and release plan • Product release backlog • Sprint planning meeting, Part 1 • Sprint planning meeting, Part 2 • Sprint backlog • Sprint iterations (time-boxed) • Sprint review meeting • Sprint retrospective meeting	APPROXIMATELY EQUIVALENT TO	• Schedule development • Schedule control

There's supplemental information in the following sections of the appendix:

Communications Management Comparison   Cost Management Comparison

Human Resources/Team Management Comparison

Integration Management Comparison   Quality Management Comparison

Risk Management Comparison   Scope Management Comparison

User Stories—Techniques for Gathering Requirements

In Scrum, user stories act as requirements. Each story represents a portion of business value that a team can deliver in an iteration. A common format is: "As a (role), I want (goal) so that I can (reason)." For example: "As a customer, I want shopping cart functionality so that I can buy items online." **4A: Product Release Backlog with User Stories and Prioritized** informs how user stories are recorded in the product release backlog. Techniques for gathering user stories are covered here.

The following methods can help the Product Owner gather material for user stories:

- Interviews

    - Ask a diverse group of users—or anticipated users if the product/service does not yet exist—open-ended questions containing "how" or "why"

        - For example: "How would you pair this device with your iPhone?"

- Observation

    - Watch people using the product/service

- Prototyping

    - Use tools such as sticky notes, PowerPoint, and wireframes to illustrate ideas, show preliminary versions of the product, and facilitate discussions

- Surveys

    - Employ surveys where the Product Owner verbally asks respondents pre-determined questions

    - Or use questionnaires where items are presented via forms (online or in hard copy format)

- Workshops

    - This is a type of brainstorming where the group identifies as many user story ideas as possible

        - To support getting a high quantity of ideas, it is suggested that participants should not agree/disagree with or assess items during the workshop

For more on user stories, see **4A: Product Release Backlog with User Stories and Prioritized**.

Velocity

Velocity is a simple but powerful method for measuring the rate at which Scrum teams deliver business value. To calculate velocity, add up the estimates (usually in story points) of the features, user stories, requirements or other backlog items completed in an iteration. Only work completed per the definition of done counts.

Velocity, Actual

- Actual velocity is the sum of the team's delivery of completed work during an iteration, usually measured in story points

    o Example 1

        ▪ A team completed work on three out of three stories in a sprint:

            • Completed story "A" had 3 points

            • Completed story "B" had 5 points

            • Completed story "C" had 8 points

        ▪ The sum of the three completed stories is 16, so the velocity is 16

    o Example 2

        ▪ A team completed work on two out of three stories in a sprint:

            • Completed story "X" had 2 points

            • Completed story "Y" had 5 points

            • Incomplete story "Z" had 5 points

        ▪ Only completed stories count; the sum of the two completed stories is 7, so the velocity is 7

- Velocity values may fluctuate from iteration to iteration, but the values often stabilize for teams after they've completed between three and six sprints

Velocity, Planned

- Planned velocity is sometimes called estimated velocity or ideal velocity

- This is the historical velocity for the team

- If the team has not done any iterations before, there is no historical data, and planned velocity does not yet apply

- If there is historical data:

    o Sum all the velocity values and divide by the number of iterations to obtain the mean average

    o Use that value as the planned velocity

- Using a simple method like the one listed above is advised, especially when starting out with Agile Scrum

- Some organizations use alternatives, such as:

    o Three-point moving average

        ▪ If the team has completed four or more iterations, sum the velocity values from the three most recent iterations and divide by three to obtain a current three-point moving average

    o Trimmed mean average

        ▪ If the team has done five or more iterations, remove the highest and lowest values and then average the remaining ones to obtain a trimmed mean average

    o Median average

        ▪ The median is the middle value of a data set

        ▪ If the list of velocity values is odd, simply take the one in the middle

            • Odd list (as it has seven values): 15, 16, 18, 19, 21, 29, 30

                – Median is 19 (the value in the middle)

        ▪ If the list of velocity values is even, take the average of the two middlemost values

            • Even list (as it has six values): 16, 18, 19, 21, 29, 31

                – Median is 20 (the average of 19 and 21)

There's supplemental information in the following sections of the appendix:

Metrics   Select Resources

When to Use Agile Scrum

Shifting customer needs are common in today's marketplace. Businesses must be adaptive and responsive to change while delivering an exceptional customer experience to be competitive. Traditional development and delivery frameworks such as waterfall are often ineffective. In contrast, Scrum is a value-driven agile approach which incorporates adjustments based on regular and repeated customer and stakeholder feedback. And Scrum's built-in rapid response to change leads to substantial benefits such as fast time-to-market, higher satisfaction, and continuous improvement—which supports innovation and drives competitive advantage.

Agile Scrum and traditional methods (such as waterfall) share planning followed by execution, monitoring, and controlling. However, there are distinctions. Key differences are highlighted below.

- Scope and requirements

    o Use Scrum when the scope and requirements are not clearly defined or frequently change

    o Use traditional methods when the scope and requirements are clearly defined and rarely change

- Process

    o Use Scrum when iteration and numerous cycles can be employed

    o Use traditional methods when the process is more long-term, and iteration does not make sense

- Success

    o Use Scrum when success is primarily measured by customer satisfaction

    o Use traditional methods when success is chiefly measured by the parameters of time, cost, and scope

There's supplemental information in the following section of the appendix:

Project Assessment Tool

Select Resources

Organizations

- Agile Alliance® — https://www.agilealliance.org

- Exceptional PPM and PMO Solutions™ — http://exceptional-pmo.com

- Project Management Institute® — http://www.pmi.org

- Rego Consulting™ — http://regoconsulting.com

- Scrum Alliance® — https://www.scrumalliance.org

- Scrum Inc.® — https://www.scruminc.com

- Scrum.org — https://www.scrum.org

Publications

- See the **Bibliography**

Blogs

- Agile Alliance® — https://www.agilealliance.org/community/blog

- Agile Scrum Guide — https://agilescrumguide.com/blog

- IndioBlue — https://www.indigoblue.co.uk/blog

- Mountain Goat Software — https://www.mountaingoatsoftware.com/blog

- ProjectManagement.com™ — https://www.projectmanagement.com/blogs

- Scrum Alliance® — https://www.scrumalliance.org/community/articles

Podcasts

- Agile for Humans™ — https://ryanripley.com/agile-for-humans

- Agile in 3 Minutes — https://agilein3minut.es

- PM Podcast™ — http://www.project-management-podcast.com

- ProjectManagement.com™ — https://www.projectmanagement.com/podcasts

- The Agile Revolution — https://theagilerevolution.com

- This Agile Life — http://www.thisagilelife.com

Technologies

- Agile Central® by CA® Technologies — http://www.ca.com

- Agile Manager by Hewlett Packard Enterprise® — https://www.hpe.com

- Agile Product Lifecycle Management by Oracle® — https://www.oracle.com

- Agilo for Scrum® by Agilo Software® — http://www.agiloforscrum.com

- Axosoft by Axosoft — https://www.axosoft.com

- iceScrum by Kagilum SAS — https://www.icescrum.com

- Jira® by Atlassian® — https://www.atlassian.com

- Klean Scrum Poker by Greener Pastures — http://www.kleancode.com

- LeanKit® by LeanKit® — https://leankit.com

- Mingle by ThoughtWorks™ — https://www.thoughtworks.com

- PivotalTracker® by Pivotal® — http://www.pivotaltracker.com

- Planbox™ by Planbox™ — https://planbox.com

- Polarion ALM® by Siemens® — https://www.siemens.com

- Scrum Poker Cards by artArmin — http://www.artarmin.com

- ScrumSwipe by AppAdvice — https://appadvice.com

- ScrumWorks® Pro by CollabNet® — http://www.collab.net

- TargetProcess by TargetProcess — https://www.targetprocess.com

- VersionOne® by VersionOne® — https://www.versionone.com

- Visual Studio® Team Foundation Server (with agile process template) by Microsoft® — https://www.microsoft.com

🔗 Any updates to this Select Resources listing after publication of the book will be available at scottgraffius.com/agile-scrum-bonus-material.html.

Feedback, Email, Social, Website, Blog, and Bonus Content

I hope that you enjoyed this quick start guide on Agile Scrum and that it brings value to your career and your organization. Please take a moment to write a review on Amazon. Thank you in advance.

— Scott M. Graffius

If you'd like, you can share your comments or suggestions via an online feedback form, communicate via email, follow on social, visit the website, read the blog, and access the online bonus content.

	Online feedback	http://scottgraffius.com/feedback.html
	Email	hello@AgileScrumGuide.com
	Twitter	https://twitter.com/AgileScrumGuide
	Facebook	https://www.facebook.com/AgileScrumGuide
	Instagram	https://www.instagram.com/AgileScrumGuide
	Website	https://AgileScrumGuide.com/
	Blog	https://AgileScrumGuide.com/blog
	Bonus content	http://scottgraffius.com/agile-scrum-bonus-material.html

Index

T

Printed in Great Britain
by Amazon

27505559R00091